A Photographic Atlas
for the
Biology Laboratory

THIRD EDITION

Kent M. Van De Graaff

John L. Crawley

Morton Publishing Company
925 W. Kenyon, Unit 12
Englewood, Colorado 80110

*To our teachers, colleagues, friends, and students
who share with us a mutual love for biology.*

Preface

Biology is an exciting, dynamic, and challenging science. It is the study of life. Students are fortunate to be living at a time when insights and discoveries in almost all aspects of biology are occurring at a very rapid pace. Much of the knowledge learned in a biology course has application in improving humanity and the quality of life. An understanding of basic biology is essential in establishing a secure foundation for more advanced courses in the life sciences or health sciences. The principles learned in a basic biology course will be of immeasurable value in dealing with the ecological and bioethical problems currently facing us.

Biology is a visually oriented science. The third edition of *A Photographic Atlas for the Biology Laboratory* is designed to provide you with quality photographs of organisms, similar to those you may have the opportunity to observe in a biology laboratory. It is designed to accompany any biology text or laboratory manual you may be using in the classroom. In certain courses, this photographic atlas could serve as the laboratory manual.

An objective of this atlas is to provide you with a balanced visual representation of the five kingdoms of biological organisms. Great care has been taken to construct completely labeled, informative figures that are depicted clearly and accurately. Because this edition of *A Photographic Atlas for the Biology Laboratory* is presented in full color, the specimens depicted in this atlas appear as you see them in the biology laboratory. The terminology used in this atlas is that found in the more commonly used college biology texts.

Several dissections of plants and invertebrate and vertebrate animals were completed and photographed in the preparation of this atlas. These images are included for those students who have the opportunity to do similar dissections as part of their laboratory requirement. In addition, photographs of a sheep heart dissection are included in this atlas.

Chapter 8 of this atlas is devoted to the biology of the human organism, which is emphasized in many biology textbooks and courses. In this chapter, you are provided with a complete set of photographs for each of the human body systems. Human cadavers have been carefully dissected and photographed to clearly depict each of the principal organs from each of the body systems. Selected radiographs (X-rays), CT scans, and MR images depict structures from living persons and thus provide an applied dimension to this portion of the atlas.

Acknowledgments

Many professionals have assisted in the preparation of all three editions of *A Photographic Atlas for the Biology Laboratory* and have shared our enthusiasm of its value for students of biology. We are especially appreciative of H. Blaine Furniss, Kyle M. Van De Graaff, and Drs. Samuel R. Rushforth, Douglas Henderson, George J. Wilder, William Ormerod, Wilford M. Hess, William B. Winborn, Neil A. Harriman, Michael J. Leboffe, Burton E. Pierce, Dawn M. Gatherum, Ferron L. Andersen, and G. R. Roberts for their help in obtaining photographs and photomicrographs.

The radiographs, CT scans, and MR images have been made possible through the generosity of Gary M. Watts, MD and the Department of Radiology at Utah Valley Regional Medical Center. Ryan L. Van De Graaff spent many hours in assisting us with dissections and photography. Several other students were very helpful in performing dissections. We gratefully acknowledge the assistance of Richard R. Rasmussen, Nathan A. Jacobson, Brad A. Chadwell, Edward R. Mathias, Hendrik W. Ombach, Sandra E. Sephton, Michelle Kidder, Bradley C. Arnold, and Michael K. Visick. Dr. Ron A. Meyers also assisted in dissections. We are grateful to Matt D. Burton, Brian C. Watts and Media Pub. Inc., for their professional talent and dedicated efforts to achieve the best possible page layout for this atlas. Christopher H. Creek and Roger Clarke rendered line art throughout the book. We appreciate their efforts.

We appreciate the suggestions from users of the first and second editions. Neil A. Harriman of the University of Wisconsin—Oshkosh, Delmar Vander Zee of Dordt College, Michael J. Shively of Utah Valley State College, David Voth of Metro State Community College, Steve Van Horn of University of Wisconsin—Stevens Point, and Edward Weiss of Christopher Newport University were especially helpful reviewers of the second edition of this atlas. Others who offered valuable input include Frank W. Ewers, Patrick F. Fields, Dale M. J. Mueller, Hohn Taylor, Brian Speer, Lawrence Virkaitis, Cecile Bochmer, and Anne S. Viscomi.

We are indebted to Douglas Morton and the personnel at Morton Publishing Company for the opportunity, encouragement, and support to prepare the third edition of this atlas.

Table of Contents

A Photographic Atlas for the Biology Laboratory

Cells and Tissues

All organisms are comprised of one or more cells. *Cells* are the basic structural and functional units of organisms. A cell is a minute, membrane-enclosed, protoplasmic mass consisting of chromosomes surrounded by cytoplasm. Specific organelles are contained in the cytoplasm that function independently but in coordination with one another. Prokaryotic cells (fig. 1.1) and eukaryotic cells (figs. 1.3 and 1.16) are the two basic types.

Prokaryotic cells lack a membrane-bound nucleus, instead containing a single strand of *nucleic acid*. These cells contain few organelles. A rigid or semi-rigid *cell wall* provides shape to the cell outside the *cell (plasma) membrane*. Bacteria are examples of prokaryotic, single-celled organisms.

Eukaryotic cells contain a true *nucleus* with multiple chromosomes, have several types of specialized organelles, and have a differentially permeable cell membrane. Organisms comprised of eukaryotic cells include protozoa, fungi, algae, plants, and invertebrate and vertebrate animals.

Plant cells differ in some ways from other eukaryotic cells in that their cell walls contain *cellulose* for stiffness (fig. 1.3). Plant cells also contain vacuoles for water storage and membrane-bound *chloroplasts* with photosynthetic pigments for photosynthesis.

The *nucleus* is the large, spheroid body within the eukaryotic cell that contains the genetic material of the cell. The nucleus is enclosed by a double membrane called the *nuclear membrane*, or *nuclear envelope*. The *nucleolus* is a dense, nonmembranous body composed of protein and RNA molecules. The chromatin are fibers of protein and DNA molecules. Prior to cellular division, the chromatin shortens and coils into rod-shaped *chromosomes*. Chromosomes consist of DNA and structural proteins called *histones*.

The *cytoplasm* of the eukaryotic cell is the medium between the nuclear membrane and the cell membrane. *Organelles* are small membrane-bound structures within the cytoplasm. The cellular functions carried out by organelles are referred to as *metabolism*. The structure and function of the nucleus and principal organelles are listed in table 1.1. In order for cells to remain alive, metabolize, and maintain homeostasis, they must have access to nutrients and respiratory gases, be able to eliminate wastes, and be in a constant, protective environment.

The *cell membrane* is composed of phospholipid, protein, and carbohydrate molecules. The cell membrane gives form to a cell and controls the passage of material into and out of a cell. More specifically, the proteins in the cell membrane provide:

1. structural support;
2. a mechanism of molecule transport across the membrane;
3. enzymatic control of chemical reactions;
4. receptors for hormones and other regulatory molecules; and
5. cellular markers (antigens), which identify the blood and tissue type.

The carbohydrate molecules:

1. repel negative objects due to their negative charge;
2. act as receptors for hormones and other regulatory molecules;
3. form specific cell markers which enable like cells to attach and aggregate into tissues; and
4. enter into immune reactions.

Tissues are groups of similar cells that perform specific functions (see fig. 1.9). A flowering plant, for example, is composed of three tissue systems:

1. the *ground tissue system*, providing support, regeneration, respiration, photosynthesis, and storage;
2. the *vascular tissue system*, providing conduction passageways through the plant; and
3. the *dermal tissue system*, providing protection to the plant.

The tissues of the body of a multicellular animal are classified into four principal types (see fig. 1.34):

1. *epithelial tissue* covers body and organ surfaces, lines body cavities and lumina (hollow portions of body tubes), and forms various glands;
2. *connective tissue* binds, supports, and protects body parts;
3. muscle tissue contracts to produce movements; and
4. nervous tissue initiates and transmits nerve impulses.

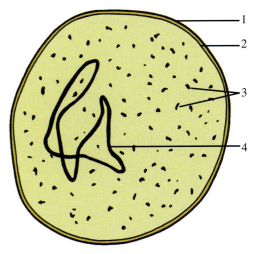

Figure 1.1 A prokaryotic cell.

1. Cell wall
2. Cell (plasma) membrane
3. Ribosomes
4. Circular molecule of DNA

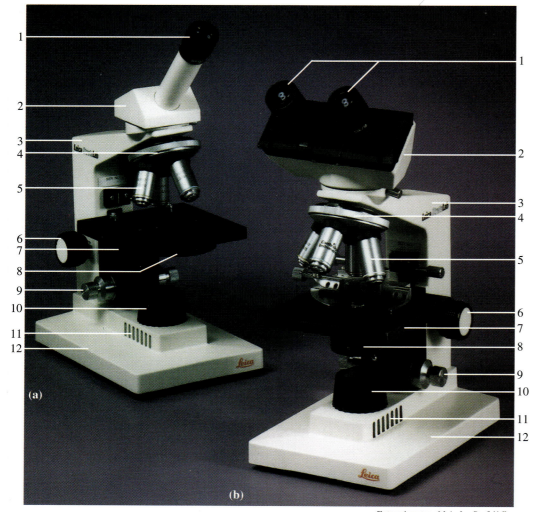

Figure 1.2 (a) A compound monocular microscope and (b) a compound binocular microscope.

1. Eyepiece(s)
2. Binocular body
3. Arm
4. Nosepiece
5. Objective
6. Focus adjustment knob
7. Fixed stage
8. Condenser
9. Stage adjustment knob
10. Collector lens with field diaphragm
11. Illuminator (inside)
12. Base

Photograph courtesy of: Leica Inc., Deerfield, IL

Table 1.1 Structure and Function of Cellular Components

Component	Structure	Function
Cell (plasma) membrane	Composed of protein and phospholipid molecules	Provides form to cell; controls passage of materials into and out of cell
Cell Wall	Cellulose fibrils	Provides structure and rigidity to plant cell
Cytoplasm	Fluid to jelly-like substance	Serves as suspending medium for organelles
Endoplasmic reticulum	Interconnecting membrane-lined channels	Provides supporting framework of cell; enables cell transport
Ribosomes	Granules of nucleic acid	Synthesize protein
Mitochondria	Double-layered sacs with cristae	Produce ATP (cellular respiration)
Golgi apparatus	Flattened membrane-lined chambers	Synthesize carbohydrates and packages molecules for secretion
Lysosomes	Membrane-surrounded sacs of enzymes	Digest foreign molecules and worn cells
Centrosome	Mass of two rodlike centrioles	Organizes spindle fibers and assists mitosis
Vacuoles	Membranous sacs	Store and secrete substances within the cytoplasm
Fibrils and microtubules	Protein strands	Support cytoplasm and transport materials
Cilia and flagella	Cytoplasmic extensions from cells	Move particles along cell surface, or move cell
Nucleus	Nuclear membrane, nucleolus, and chromatin (DNA)	Directs cell activity; forms ribosomes
Chloroplast	Inner (grana) membrane within outer membrane	

Plant cells and tissues

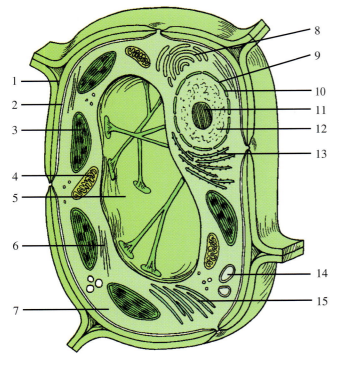

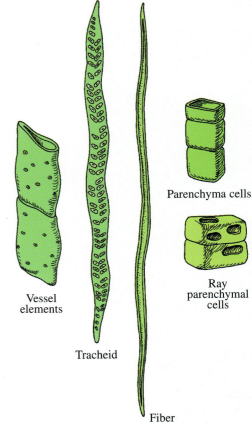

Figure 1.3 A typical plant cell.

1. Cell wall
2. Cell (plasma) membrane
3. Chloroplast
4. Mitochondrion
5. Vacuole
6. Microfilament
7. Cytoplasm
8. Golgi apparatus
9. Nuclear pore
10. Nuclear membrane (envelope)
11. Nucleolus
12. Chromatin
13. Rough endoplasmic reticulum
14. Vesicle
15. Smooth endoplasmic reticulum

Vessel elements

Tracheid

Fiber

Parenchyma cells

Ray parenchymal cells

Figure 1.4 Examples of plant cells.

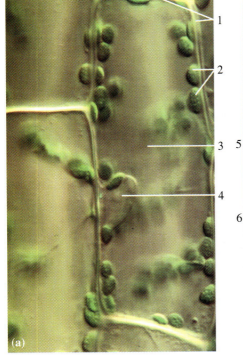

(a)   (X430)

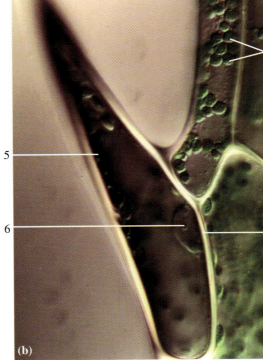

(b)   (X430)

Figure 1.5 Live *Elodea* leaf cells (a) photographed at the center of the leaf and (b) at the edge of the leaf.

1. Cell wall
2. Chloroplasts
3. Vacuole
4. Nucleus
5. Spine-shaped cell on exposed edge of leaf
6. Nucleus
7. Chloroplasts
8. Cell wall

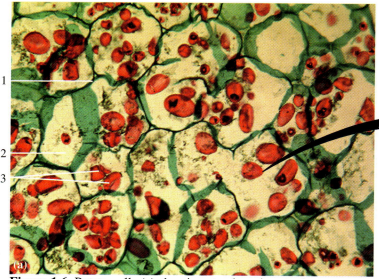

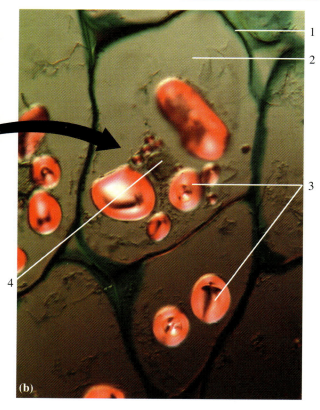

Figure 1.6 Potato cells (a) showing starch grains at a low magnification of 430, and (b) at a high magnification of 1,000. Food is stored as starch in potato cells, which is deposited in organelles called amyloplasts.

1. Cell wall
2. Cytoplasm
3. Starch grains
4. Nucleus

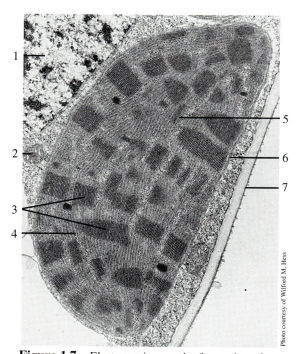

Photo courtesy of Wilford M. Hess

Figure 1.7   Electron micrograph of a portion of a sugar cane leaf cell.

1. Nucleus
2. Mitochondrion
3. Grana
4. Thylakoid membrane
5. Stroma
6. Chloroplast membrane
7. Cell wall

Photo courtesy of Wilford M. Hess

Figure 1.8   Barley smut spore, fractured through the middle of the cell.

1. Nucleus
2. Vacuole
3. Cell wall
4. Cell membrane
5. Mitochondrion
6. Nuclear pores

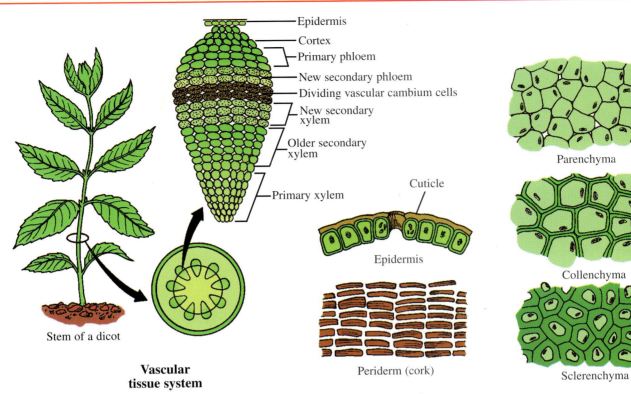

Stem of a dicot

Vascular tissue system

Figure 1.9 Examples of plant tissues.

Cuticle

Epidermis

Periderm (cork)

Dermal tissue system

Parenchyma

Collenchyma

Sclerenchyma

Ground tissue system

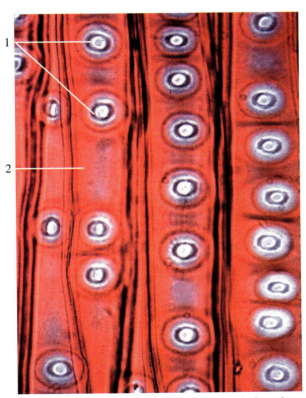

Figure 1.10 A Longitudinal section through the xylem of a pine (*Pinus*) showing tracheid cells with prominent bordered pits.

1. Bordered pits   2. Tracheid cell

Figure 1.11 A Longitudinal section through the xylem of a squash stem (*Cucurbita maxima*). The vessel elements shown here have several different patterns of wall thickenings.

1. Vessel elements
2. Parenchyma

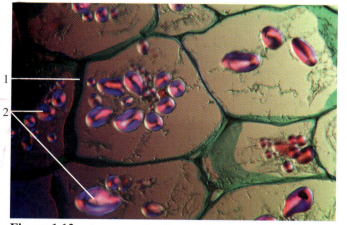

Figure 1.12   A cross section through the stem (tuber) of a potato (*Solanum tuberosum*) showing parenchyma cells containing numerous starch grains.

1. Cell wall   2. Starch grains

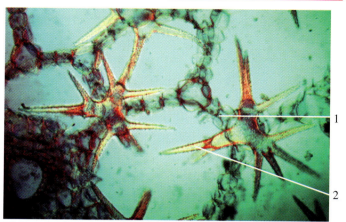

Figure 1.13   An asterosclereid in the petiole of a pondlily (*Nuphar*).

1. Parenchyma cell   2. Asterosclereid

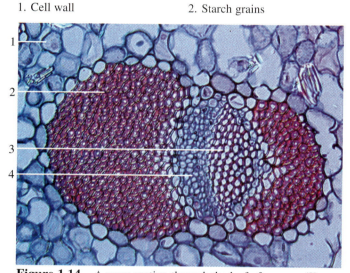

Figure 1.14   A cross section through the leaf of a yucca (*Yucca brevifolia*) showing a vascular bundle (vein). Notice the prominent sclerenchyma tissue forming caps on both sides of the bundle.

1. Leaf parenchyma   3. Xylem
2. Leaf sclerenchyma   4. Phloem
   (bundle sheath)

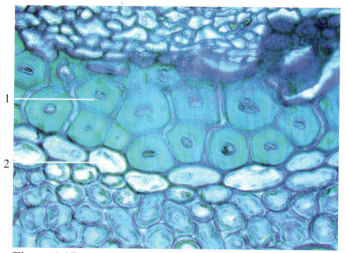

Figure 1.15   A cross section through the stem of flax (*Linum*). Notice the thick-walled fibers as compared to the thin-walled parenchyma cells.

1. Fibers   2. Parenchyma cell

Animal cells and tissues

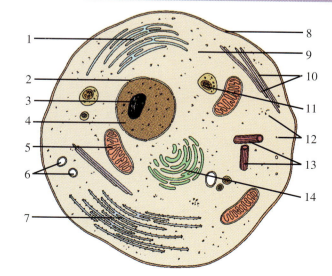

Figure 1.16 A typical animal cell.

1. Smooth endoplasmic   8. Cell membrane
   reticulum   9. Cytoplasm
2. Nuclear membrane   10. Microtubules
3. Nucleolus   11. Lysosome
4. Nucleoplasm   12. Ribosomes
5. Mitochondrion   13. Centrioles
6. Vesicles   14. Golgi apparatus
7. Rough endoplasmic
   reticulum

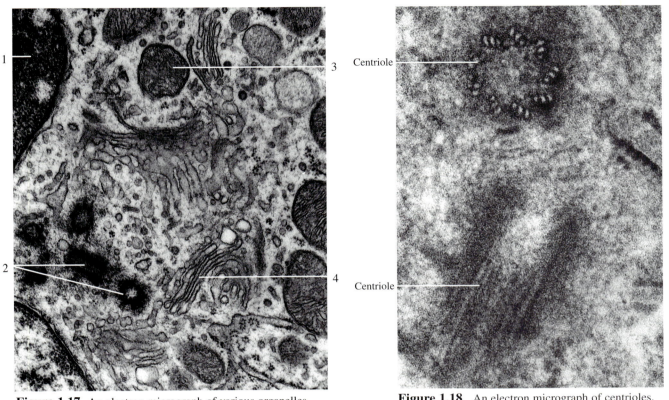

Figure 1.17   An electron micrograph of various organelles.

1. Nucleus   3. Mitochondrion
2. Centrioles   4. Golgi apparatus

Figure 1.18   An electron micrograph of centrioles. The centrioles are positioned at right angles to one another.

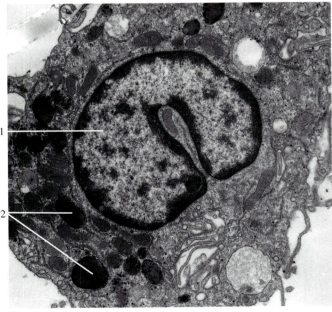

Figure 1.19   An electron micrograph of lysosomes.

1. Nucleus   2. Lysosomes

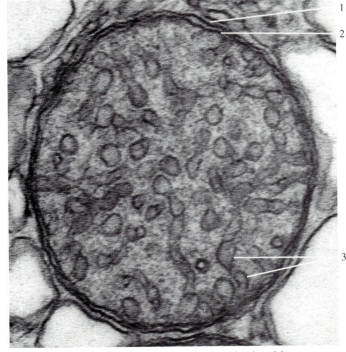

Figure 1.20   An electron micrograph of a mitochondrion.

1. Outer membrane   3. Inner membranes
2. Crista

Photo courtesy of Scott C. Miller

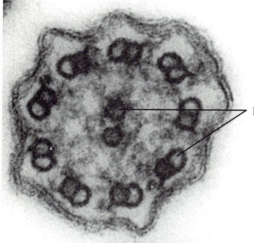

Figure 1.21 An electron micrograph of cilia showing the characteristic "9 + 2" arrangement of microtubules in the cross sections.

1. Microtubules

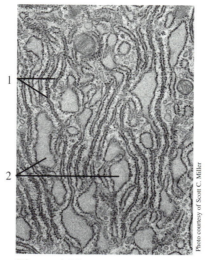

Photo courtesy of Scott C. Miller

Figure 1.22 An electron micrograph of rough endoplasmic reticulum.

1. Ribosomes   2. Cisternae

Photo courtesy of Scott C. Miller

Figure 1.23 Rough endoplasmic reticulum secreting collagenous filaments to the outside of the cell.

1. Nucleus
2. Rough endoplasmic reticulum
3. Collagenous filaments
4. Cell membrane

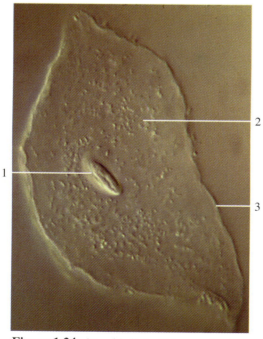

Figure 1.24 An epithelial cell from a cheek scraping.

1. Nucleus
2. Cytoplasm
3. Cell membrane

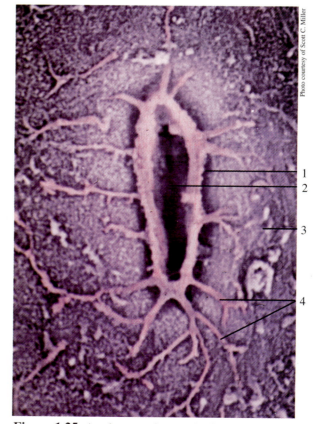

Photo courtesy of Scott C. Miller

Figure 1.25 An electron micrograph of an osteocyte in cortical bone matrix.

1. Lacuna
2. Osteocyte
3. Bone matrix
4. Canaliculi

Photo courtesy of Scott C. Miller

Figure 1.26 An electron micrograph of a skeletal muscle myofibril, showing the striations.

1. Mitochondria   5. T-tubule
2. Z line   6. Sarcoplasmic reticulum
3. I band   7. H band
4. A band   8. Sacromere

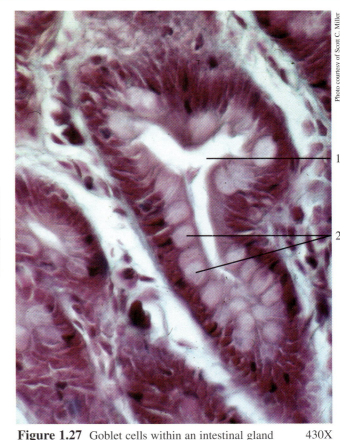

Photo courtesy of Scott C. Miller

Figure 1.27 Goblet cells within an intestinal gland (crypt of Lieberkühn) of small intestine.   430X

1. Lumen of gland
2. Goblet cells

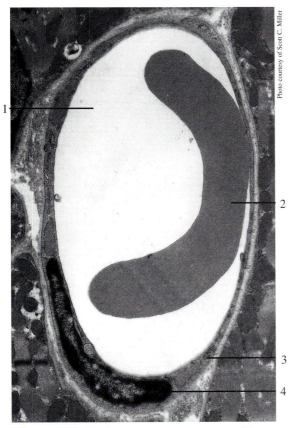

Photo courtesy of Scott C. Miller

Figure 1.28 An electron micrograph of a capillary containing an erythrocyte.

1. Lumen of capillary   3. Endothelial cell
2. Erythrocyte   4. Nucleus of endothelial cell

Photo courtesy of Scott C. Miller

Figure 1.29 An electron micrograph of an erythrocyte (red blood cell).

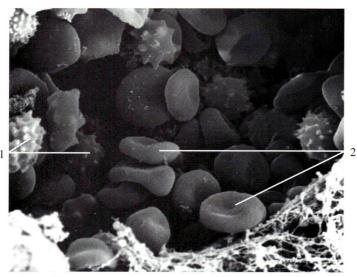

Figure 1.30 An electron micrograph of blood cells in the lumen of a blood vessel.

1. Leukocytes
2. Erythrocytes

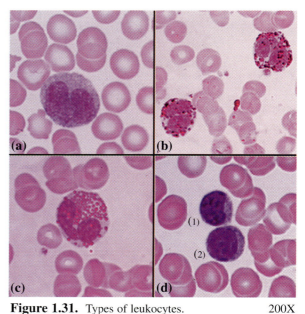

Figure 1.31. Types of leukocytes.   200X

(a) Neutrophil   (d) Monocyte (1);
(b) Basophils   lymphocyte (2)
(c) Eosinophil

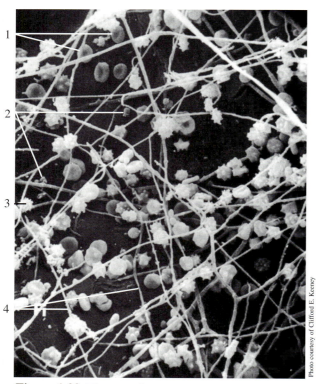

Photo courtesy of Clifford E. Keeney

Figure 1.32 Electron micrograph of a blood clot.

1. Erythrocytes   3. Thrombocytes
2. Leukocytes   4. Fibrin strand

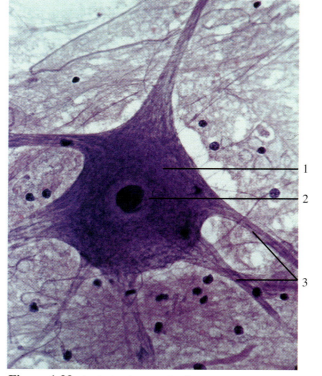

Figure 1.33 A photomicrograph of a neuron.

1. Cell body of neuron
2. Nucleus
3. Cytoplasmic extensions

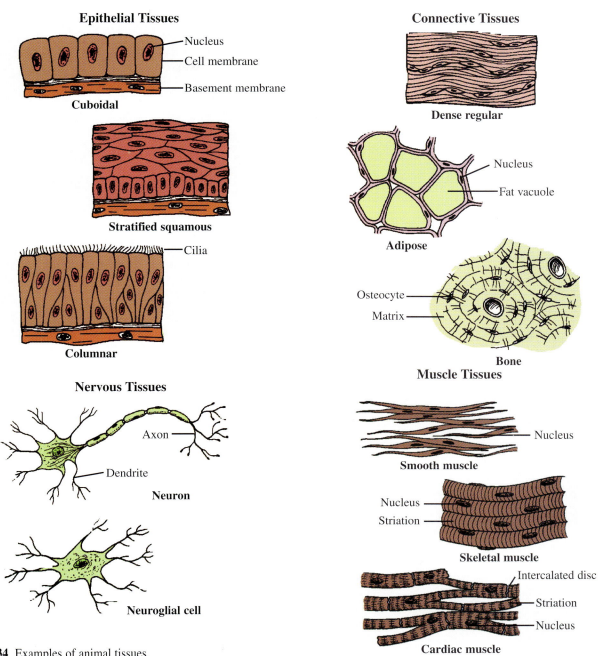

Epithelial Tissues

Nucleus
Cell membrane
Basement membrane

Cuboidal

Stratified squamous

Cilia

Columnar

Connective Tissues

Dense regular

Nucleus
Fat vacuole

Adipose

Osteocyte
Matrix

Bone

Nervous Tissues

Axon
Dendrite

Neuron

Neuroglial cell

Muscle Tissues

Nucleus

Smooth muscle

Nucleus
Striation

Skeletal muscle

Intercalated disc
Striation
Nucleus

Cardiac muscle

Figure 1.34 Examples of animal tissues.

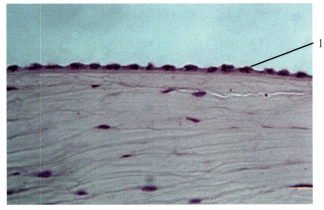

Figure 1.35 Simple squamous epithelium.   300X
1. Single layer of flattened cells

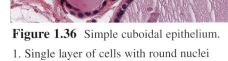

Figure 1.36 Simple cuboidal epithelium.   300X
1. Single layer of cells with round nuclei

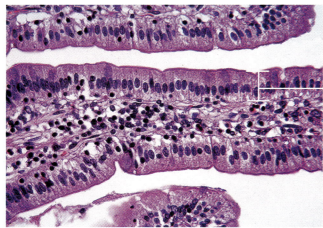

Figure 1.37 Simple columnar epithelium.   300X

1. Single layer of cells with oval nuclei

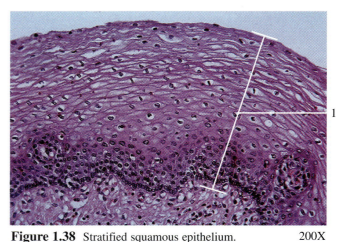

Figure 1.38 Stratified squamous epithelium.   200X

1. Multiple layers of cells, which are flattened at the

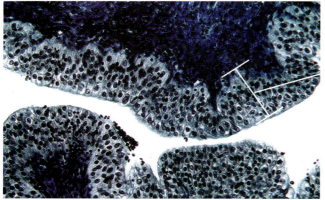

Figure 1.39 Stratified columnar epithelium.   200X

1. Cells are balloon-like at surface

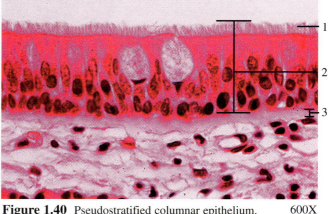

Figure 1.40 Pseudostratified columnar epithelium.   600X

1. Cilia
2. Pseudostratified columnar epithelium
3. Basement membrane

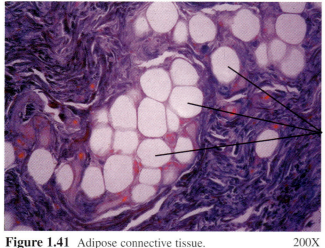

Figure 1.41 Adipose connective tissue.   200X

1. Adipocytes (adipose cells)

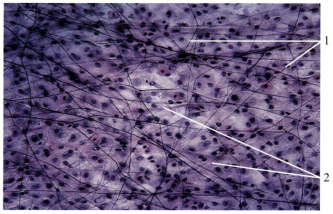

Figure 1.42 Loose connective tissue stained   200X
for elastic and collagen fibers.

1. Elastic fibers (black)
2. Collagen fibers (pink)

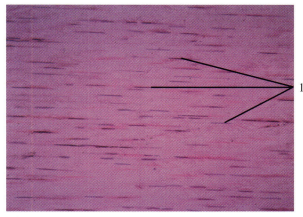

Figure 1.43 Dense regular connective tissue. 200X

1. Nuclei of fibroblasts arranged in parallel rows

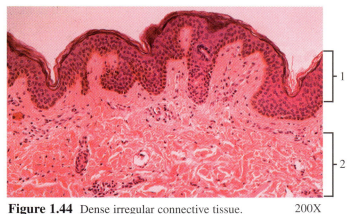

Figure 1.44 Dense irregular connective tissue. 200X

1. Epidermis
2. Dense irregular connective tissue (reticular layer of dermis)

Figure 1.45 An electron micrograph of dense irregular connective tissue. 200X

1. Collagenous fibers

Figure 1.46 Reticular connective tissue. 200X

1. Reticular fibers

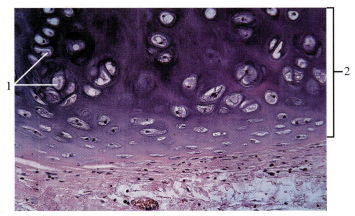

Figure 1.47 Hyaline cartilage. 200X

1. Chondrocytes
2. Hyaline cartilage

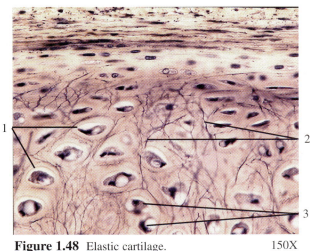

Figure 1.48 Elastic cartilage. 150X

1. Lacunae 3. Chondrocytes
2. Elastic fibers

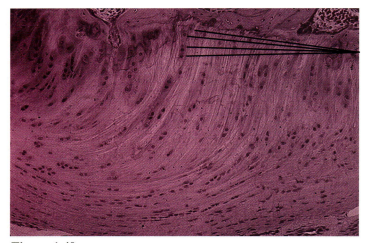

Figure 1.49 Fibrocartilage.   150X

1. Chondrocytes arranged in a row

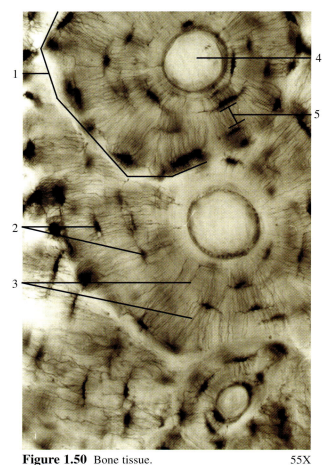

Figure 1.50 Bone tissue.   55X

1. Osteon (Haversian system)
2. Osteocytes in lacunae
3. Canaliculi
4. Central canal (Haversian canal)
5. Lamella of osteon

(a)
                                                                       200X

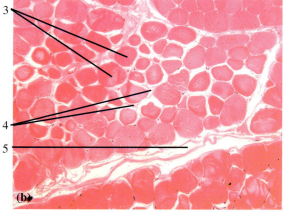

(b)

Figure 1.51 Skeletal muscle tissue.   200X
(a) A longitudinal section and (b) a cross section.

1. Nuclei   4. Endomysium
2. Striations of skeletal muscle fibers   5. Perimysium
3. Skeletal muscle fibers

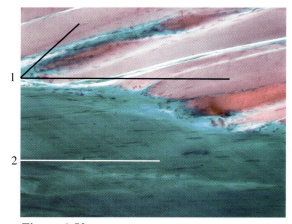

Figure 1.52 The attachment of skeletal   200X
muscle fibers to a tendon.

1. Skeletal muscle fibers
2. Tendon composed of dense regular connective tissue

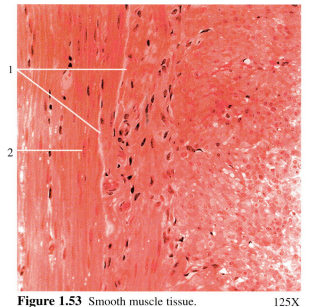

Figure 1.53 Smooth muscle tissue.   125X

1. Smooth myofibers
2. Nucleus of smooth myofiber

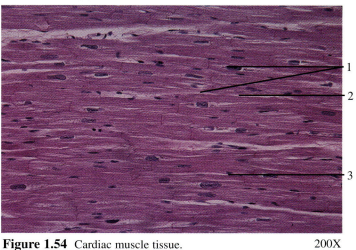

Figure 1.54 Cardiac muscle tissue.   200X

1. Intercalated discs
2. Light-staining perinuclear sarcoplasm
3. Nucleus in center of cell

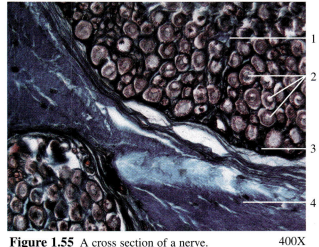

Figure 1.55 A cross section of a nerve.   400X

1. Endoneurium   3. Perineurium
2. Axons   4. Epineurium

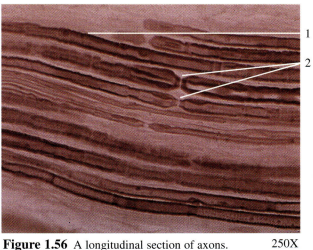

Figure 1.56 A longitudinal section of axons.   250X

1. Myelin sheath
2. Neurofibril nodes (nodes of Ranvier)

Perpetuation of Life

The term *cell cycle* refers to how a multicellular organism develops, grows, maintains and repairs body tissues. In the cell cycle, each new cell receives a complete copy of all genetic information in the parent cell, and the cytoplasmic substances and organelles to carry out hereditary instructions.

The animal cell cycle (see fig. 2.5) is divided into: 1) interphase, which includes G1, S, and G2 phases; and 2) mitosis, which includes prophase, metaphase, anaphase, and telophase. *Interphase* is the interval between successive cell divisions during which the cell is metabolizing and the chromosomes are directing RNA synthesis. The *G1 phase* is the first growth phase, the *S phase* is when DNA is replicated, and the *G2 phase* is the second growth phase. *Mitosis* (also karyokinesis) is the division of the nuclear parts of a cell to form two diploid daughter nuclei.

Like the animal cell cycle, the plant cell cycle consists of growth, synthesis, mitosis, and cytokinesis. *Growth* is the increase in cellular mass as the result of metabolism; *synthesis* is the production of DNA and RNA to regulate cellular activity; mitosis is the splitting of the nucleus and the equal separation of the chromatids; and cytokinesis is the division of the cytoplasm that accompanies mitosis.

Unlike animal cells, plant cells have a rigid cell wall that does not cleave during cytokinesis. Instead, a new cell wall is constructed between the daughter cells. Furthermore, many land plants do not have centrioles for the attachment of spindles. The microtubules in these plants form a barrel-shaped anastral spindle at each pole. Mitosis and cytokinesis in plants occurs in basically the same sequence as these processes in animal cells.

Asexual reproduction is propagation without sex; that is, the production of new individuals by processes that do not involve *gametes* (sex cells). Asexual reproduction occurs in a variety of microorganisms, fungi, plants, and animals, wherein a single parent produces offspring with characteristics identical to itself. Asexual reproduction is not dependent on the presence of other individuals. No egg or sperm is required. In asexual reproduction, all the offspring are genetically identical (except for mutants). Types of asexual reproduction include:

1. *fission*—a single cell divides to form two separate cells (bacteria, protozoans, and other one-celled organisms);
2. *sporulation*—multiple fission, many cells are formed and join together in a cystlike structure (protozoans and fungi);
3. *budding*—buds develop organs like the parent and then detach themselves (hydras, yeast, certain plants); and
4. *fragmentation*—organisms break into two or more parts, and each part is capable of becoming a complete organism (algae, flatworms, echinoderms).

Sexual reproduction is propagation of new organisms through the union of genetic material from two parents. Sexual reproduction usually involves the fusion of haploid gametes (such as sperm and egg cells) during fertilization to form a zygote.

The major biological difference between sexual and asexual reproduction is that sexual reproduction produces genetic variation in the offspring. The combining of genetic material from the gametes produces offspring that are different from either parent and contain new combinations of characteristics. This may increase the ability of the species to survive environmental changes or to reproduce in new habitats. The only genetic variation that can arise in asexual reproduction comes from mutations.

Figure 2.1 Types of asexual reproduction.

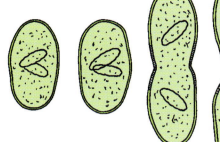

Fission

A single cell divides forming two separate cells. Fission occurs in bacteria, protozoans, and other single-celled organisms.

Budding

A plant produces external stems, or runners. Budding occurs in a number of flowing plants, such as strawberries.

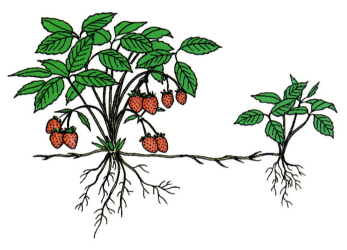

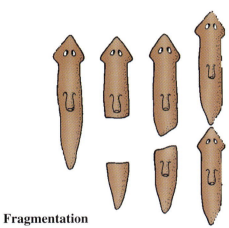

Fragmentation

An organism breaks into two or more parts, each capable of becoming a complete organism. Fragmentation occurs in flatworms and echinoderms.

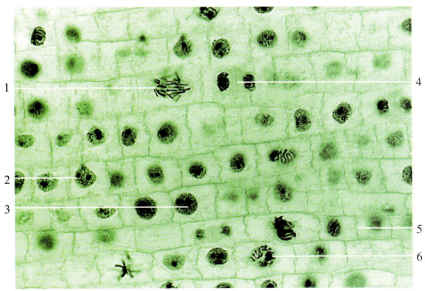

Figure 2.2 Cells in various stages of mitosis from an onion (*Allium*) root tip.

200X

1. Anaphase
2. Interphase
3. Early prophase
4. Telophase
5. Cell wall
6. Late prophase

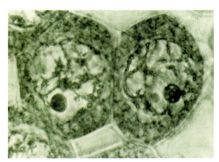

Prophase I — Each chromosome consists of two chromatids joined by a centromere. Spindle fibers extend from each centriole.

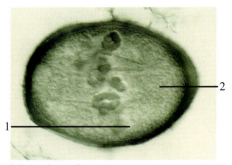

Metaphase I — The chromosomes align at the equator with their homologous partner. During this stage, called synapsis, exchange occurs between the chromosomes.

1. Chromatids at equator
2. Spindle fibers

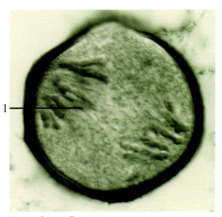

Anaphase I — No division at the centromeres occurs as the chromosomes separate, so one copy of each homologous pair of chromosomes goes to each pole.

1. Chromatids

Telophase I — The chromosomes lengthen and become less distinct. The cell wall (in some plants) forms between the forming cells.

1. Cell wall

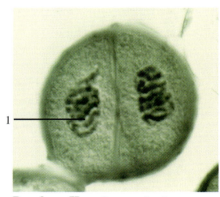

Prophase II — Once again, the chromosomes condense as in prophase I.

1. Chromatid

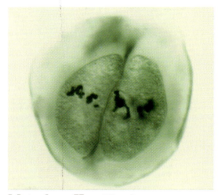

Metaphase II — The chromosomes align on the equator and the spindle fibers attach to the centromeres. This is similar to metaphase in mitosis.

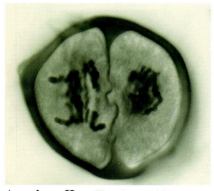

Anaphase II — The chromatids separate and each is pulled to an opposite pole.

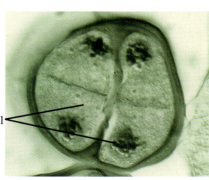

Telophase II — Cell division is complete and cell walls of four haploid cells are formed.

1. Cell walls

Figure 2.3 Stages of meiosis in lily microspore mother cells. (all 1000X)

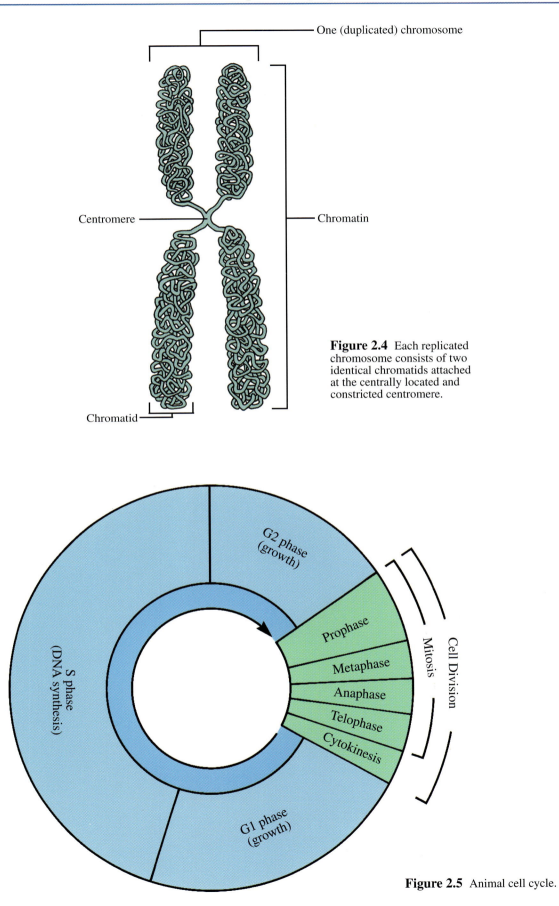

One (duplicated) chromosome

Centromere

Chromatin

Figure 2.4 Each replicated chromosome consists of two identical chromatids attached at the centrally located and constricted centromere.

Chromatid

G2 phase (growth)

Prophase

Metaphase

Anaphase

Telophase

Cytokinesis

Mitosis

Cell Division

S phase (DNA synthesis)

G1 phase (growth)

Figure 2.5 Animal cell cycle.

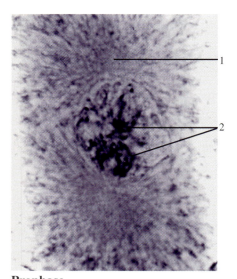

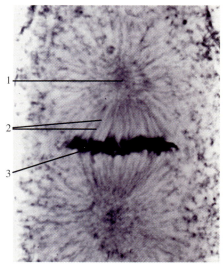

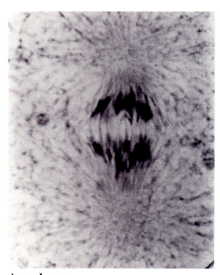

Prophase

Each chromosome consists of two chromatids jointed by a centromere. Spindle fibers extend from each centriole.

1. Centriole   2. Chromosomes

Metaphase

The chromosomes are positioned at the equator. The spindle fibers from each centriole attach to the centromeres.

1. Aster around   3. Chromosomes at
   centriole   equator
2. Spindle fibers

Anaphase

The centromeres split, and the sister chromatids separate as each is pulled to an opposite pole.

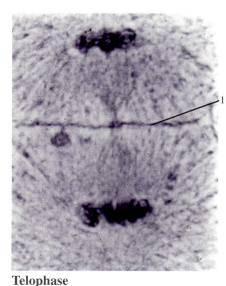

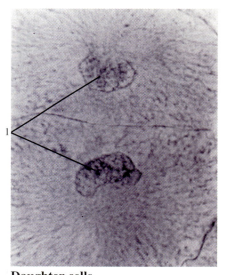

Figure 2.6 Stages of animal cell mitosis. 1000X

Telophase

The chromosomes lengthen and become less distinct. The cell membrane forms between the forming daughter cells.

1. Cell membrane

Daughter cells

The single chromosomes (former chromatids—see anaphase) continue to lengthen as the nuclear membrane reforms. Cell division is complete and the newly formed cells grow and mature.

1. Daughter nuclei

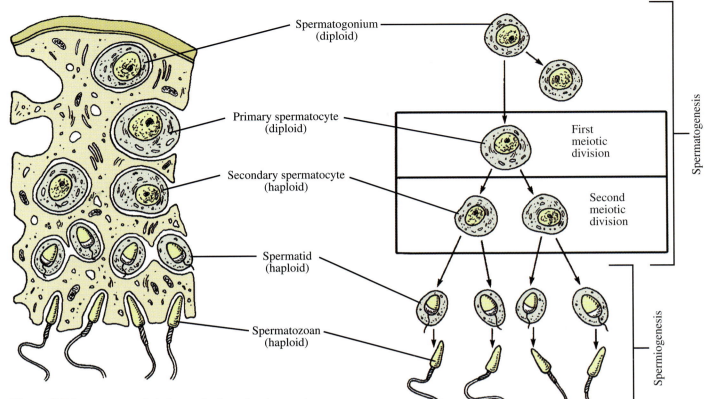

Figure 2.7 Spermatogenesis is the production of male gametes, or spermatozoa, through the process of meiosis.

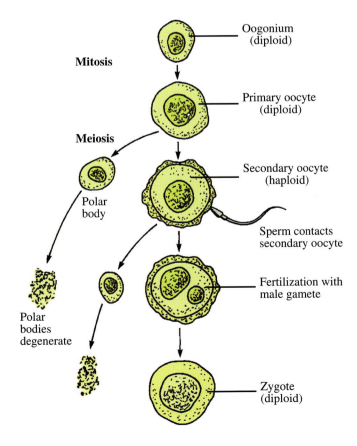

Figure 2.9 (a) An intact chicken egg and (b) a portion of the shell is removed exposing the internal structures.

1. Shell
2. Vitelline membrane
3. Yolk
4. Albumen (egg white)
5. Shell
6. Chalaza (dense albumen)
7. Air space
8. Shell membrane

Figure 2.8 Oogenesis is the production of female gametes, or ova, through the process of meiosis.

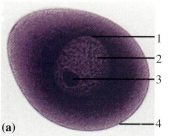

(a)

Unfertilized egg.

1. Nuclear membrane
2. Nucleus
3. Nucleolus
4. Cell membrane

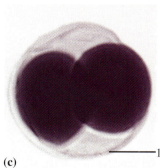

(b)

Fertilized egg.

1. Fertilization membrane

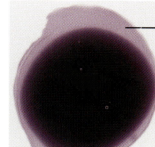

(c)

2-cell stage.

1. Fertilization membrane

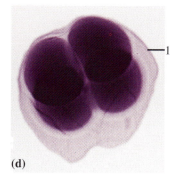

(d)

4-cell stage.

1. Fertilization membrane

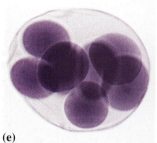

(e)

8-cell stage.

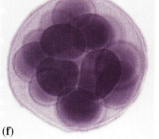

(f)

16-cell stage.

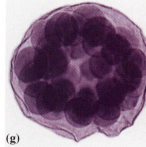

(g)

32-cell stage.

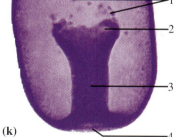

(h)

64-cell stage.

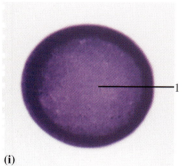

(i)

Blastula

1. Blastocoel

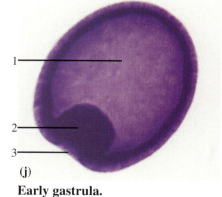

(j)

Early gastrula.

1. Blastocoel   3. Blastopore
2. Archenteron
   (gastrocoel)

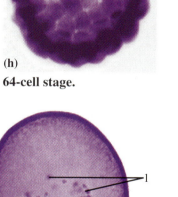

(k)

Late gastrula.

1. Mesenchyme cells   4. Blastopore
2. Coelomic sac
3. Archenteron
   (gastrocoel)

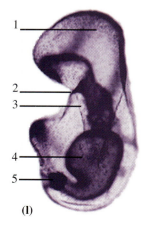

(l)

Bipinnaria larva

(side view).

1. Oral lobe
2. Mouth
3. Coelomic pouch
4. Stomach
5. Anus

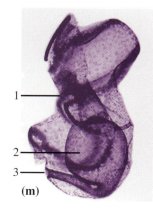

(m)

Early brachiolaria larva

(side view).

1. Mouth
2. Stomach
3. Anus

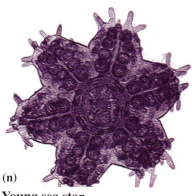

(n)

Young sea star.

Figure 2.10 Sea star development.

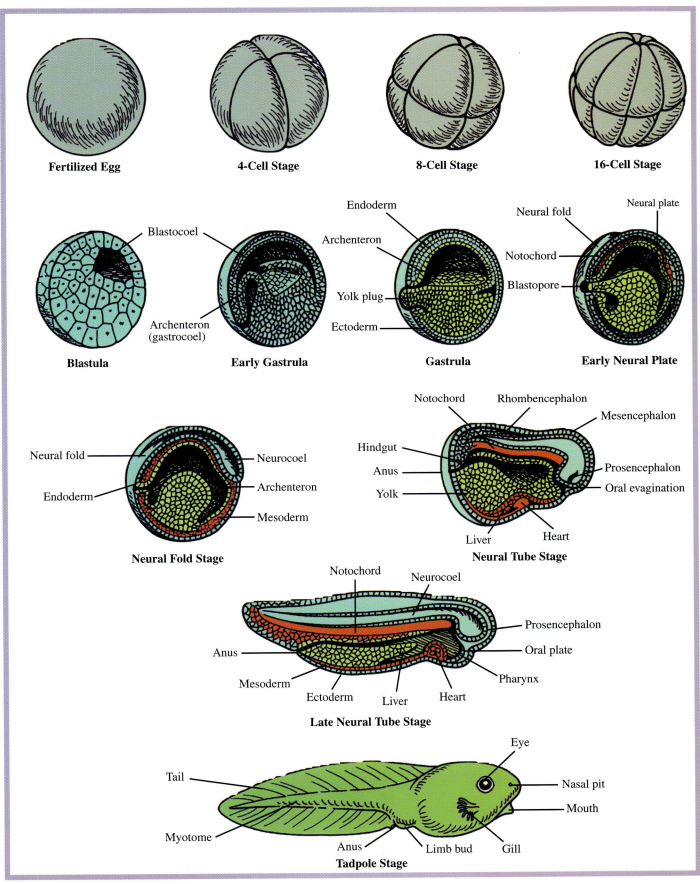

Figure 2.11 Frog development.

Kingdom Monera

3

The kingdom Monera (about 2,500 species of bacteria) contains all organisms comprised of prokaryotic cells. Prokaryotic cells were the first kinds of cells to evolve, probably about 3.5 billion years ago. The Archaebacteria and Eubacteria are the two groupings within the kingdom Monera.

Archaebacteria are adapted to a limited range of extreme conditions. The cell walls of Archaebacteria lack peptidoglycan (characteristic of Eubacteria). Archaebacteria have distinctive transfer RNAs and RNA polymerases. They include methanogens, typically found in swamps and marshes, and thermoacidophiles, found in acid hot springs and acidic soil.

Methanogens are bacteria that exist in oxygen-free environments and subsist on simple, inorganic compounds such as CO_2, acetate, and methanol. As their name implies, Methanobacteria produce methane gas as a byproduct of metabolism. These organisms are typically found in organic-rich mud and sludge, particularly that which contains fecal wastes.

The thermoacidophiles are resistant to hot temperatures and high acid concentrations. The cell membrane of these organisms contains high amounts of saturated fats, and its enzymes and other proteins are able to withstand extreme conditions without denaturation. These microscopic organisms thrive in most hot springs and hot, acid soils.

In contrast to Archaebacteria, Eubacteria are considered the "true" bacteria. Eubacteria include the Cyanobacteria (formerly known as blue-green algae) and a number of other diverse types (table 3.1). Cyanobacteria are photosynthetic bacteria that contain chlorophyll and release oxygen during photosynthesis. Some bacteria are obligate aerobes (require O_2 for metabolism) and others are facultative anaerobes (indifferent to O_2 for metabolism). Most bacteria are heterotrophic saprophytes, which secrete enzymes to break down surrounding molecules into absorbable compounds.

Bacteria range between 1 and 10 μm in width or diameter. The morphological appearance may be spiral (spirillum), spherical (coccus), or rod-shaped (bacillus). Cocci and bacilli frequently form clusters or linear filaments, and many have cilia. Relatively few species of bacteria cause infection. Hundreds of species of non-pathogenic bacteria live on the human body and within the gastrointestinal (GI) tract. Those in the GI tract constitute a person's normal gut fauna.

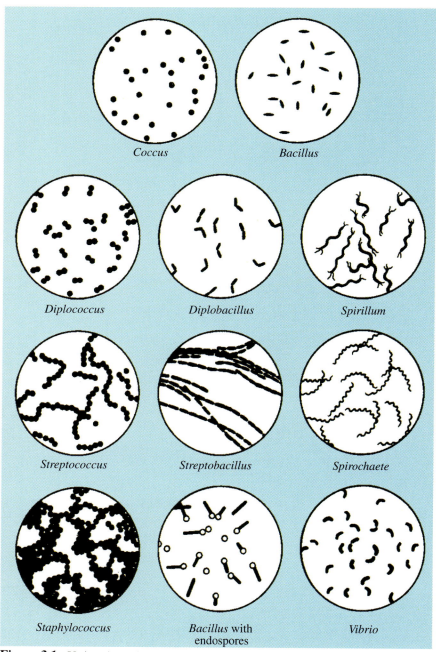

Figure 3.1 Various bacteria.

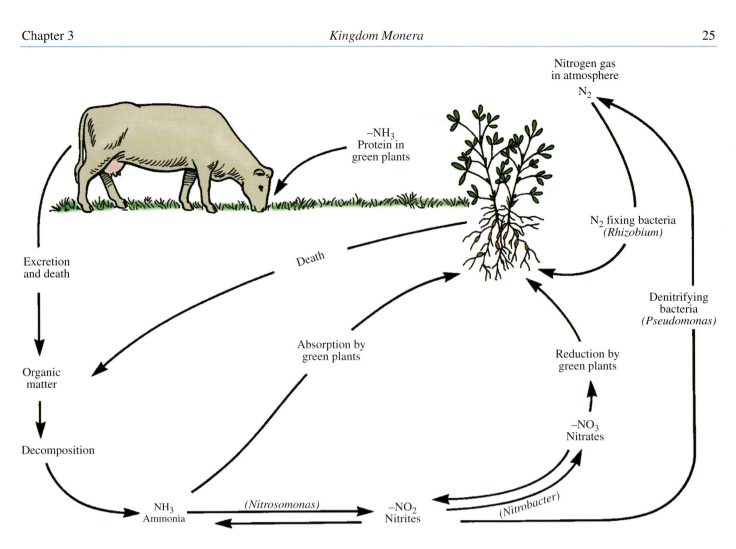

Figure 3.3 Nitrogen-fixing bacteria within the root nodules of legumes provide a usable source of nitrogen to plants.

Table 3.1 Some Representatives of the Kingdom Monera

Categories	Representative Genera
Archaebacteria	
Methanogens	*Halobacterium, Methanobacteria*
Thermoacidophiles	*Thermoplasma, Sulfobolus*
Eubacteria	
Photosynthetic bacteria Cyanobacteria	*Anabaena, Oscillatoria, Spirulina, Nostoc*
Green bacteria	*Chlorobium*
Purple bacteria	*Rhodospirillum*
Gram-negative bacteria	*Proteus, Pseudomonas, Escherichia, Rhizobium, Neisseria*
Gram-positive bacteria	*Streptococcus, Staphylococcus, Bacillus, Clostridium, Lactobacillus*
Spirochaetes	*Spirochaeta, Treponema*
Actinomycetes	*Actinomyces*
Rickettsias	*Rickettsia, Chlamydia*
Mycoplasmas	*Mycoplasma*

Figure 3.2 A *Nostoc* filament. Individual filaments secrete mucilage which forms a binding matrix.   430X
1. Filament

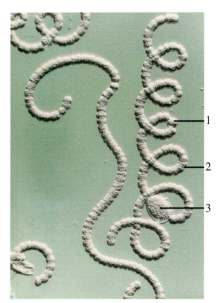

Figure 3.4   *Anabaena*   430X
filaments. This is a nitrogen-fixing
cyanobacterium. Nitrogen fixation takes
place within the heterocyst cells.

1. Heterocyst
2. Vegetative cell
3. Akinete (spore)

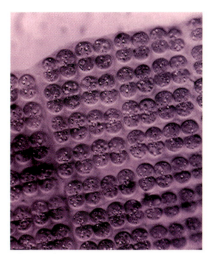

Figure 3.5   *Merismopedia,*   430X
a genus of cyanobacterium, is
characterized by flattened colonies of
cells. The cells are in a single-layer and
usually aligned into groups of two or
four.

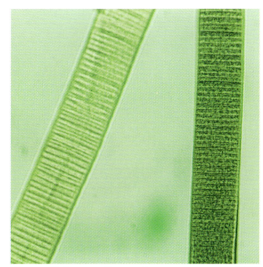

Figure 3.6   *Oscillatoria* filaments.   430X
The only way this cyanobacterium
can reproduce is through fragmentation
of a filament.

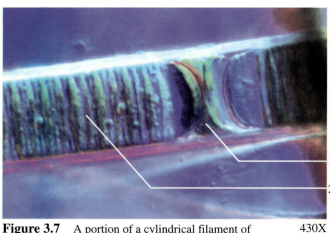

Figure 3.7   A portion of a cylindrical filament of   430X
Oscillatoria. This cyanobacterium is common in most
aquatic habitats.

1. Separation disk
2. Filament segment

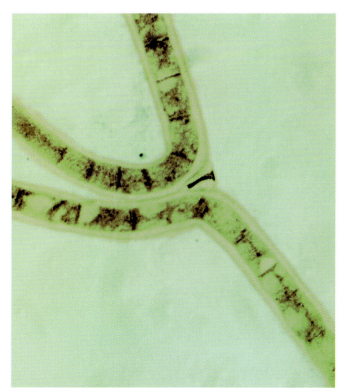

Figure 3.8   *Scytonema*, a cyanobacterium, is   430X
common on soil moistened from the spray
of a waterfall or stream. Notice the falsely-branched
filament typical of this genus.

Figure 3.9   *Stigonema*, a cyanobacterium, has true branched filaments.   500X

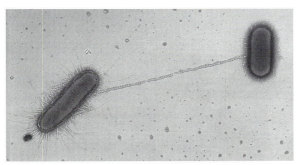

Figure 3.11   Conjugation of the bacterium *Escherichia* coli. By this process, genetic material is transferred through the conjugation tube from one cell to the other.   1700X

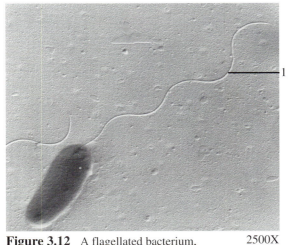

Figure 3.12   A flagellated bacterium, *Pseudomonas*.   2500X

1. Flagellum

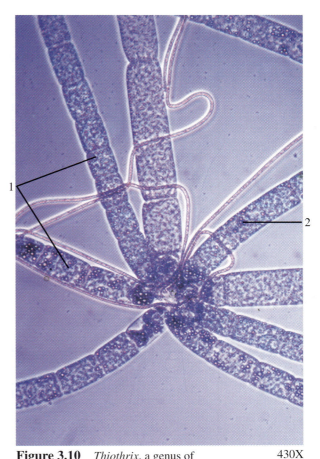

Figure 3.10   *Thiothrix*, a genus of bacteria that forms sulfur granules in its cytoplasm. These organisms obtain energy from oxidation of H_2S.   430X

1. Filaments   2. Sulfur granules

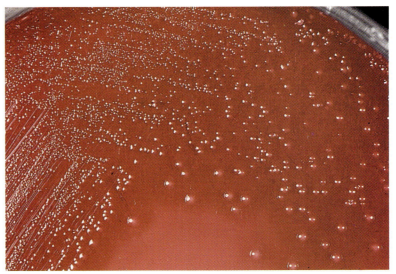

Figure 3.13   Colonies of *Streptococcus pyogenes* cultured on a nutrient agar plate. *S. pyogenes* causes strep throat and rheumatic fever.

Figure 3.14 *Bacillus megaterium.*   1000X
Bacillus is one of the few bacteria capable of producing endospores. This species of *Bacillus* generally remains in chains after it divides.

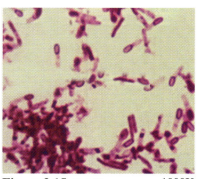

Figure 3.15 *Spirillum volutans.* Bacteria in the genus *Spirillum* are shaped like long rods twisted into rigid helices. They generally have multiple polar flagella.   1000X

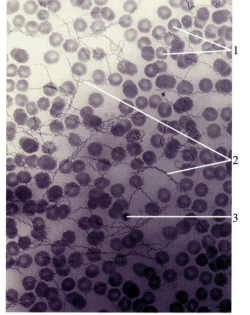

Figure 3.16 A spirochete, *Borella*   1000X
recurrentis. Spirochetes are flexible rods twisted into helical shapes. This species causes relapsing fever.

1. Red blood cells
2. Spirochete
3. White blood cells

Figure 3.17 *Streptococcus pyogenes,* the organism that causes rheumatic fever. Notice the chains of coccus-shaped bacteria.

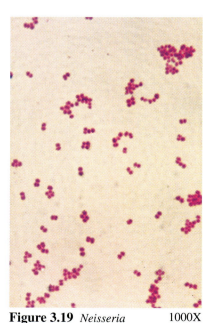

1000X

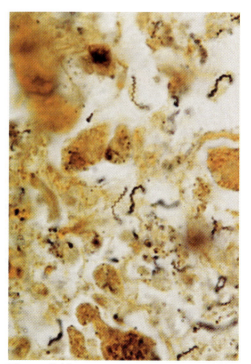

Figure 3.18 *Treponema pallidum*   1000X
is a spirochete that causes syphilis.

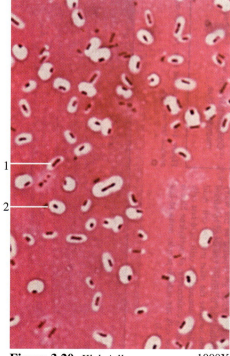

Figure 3.19 *Neisseria*   1000X
gonorrhoeae is a diplococcus that causes gonorrhea.

Figure 3.20 *Klebsiella*   1000X
pneumoniae capsules. This bacterium is able to encapsulate itself in order to make it more resistant to host defense mechanisms.
1. Cell
2. Capsule

Kingdom Protista

Most protists are unicellular, eukaryotic organisms, although some species are multicellular. Protists have a nucleus, mitochondria, chloroplasts, endoplasmic reticulae, and Golgi apparati. Protists are capable of meiosis and sexual reproduction; these processes evolved a billion or more years ago and occur in nearly all complex plants and animals.

Protists are abundant in aquatic habitats, and are important constituents of plankton. Plankton are communities of organisms that drift passively or swim slowly near the surface of ponds, lakes, and oceans. Plankton are a major source of food for other aquatic organisms. Photosynthetic protists are the primary food produces in aquatic ecosystems.

The unicellular algal protists include microscopic aquatic organisms within the phyla Chrysophyta and Pyrrhophyta. Chrysophyta are the yellow-green and the golden-brown algae, and the diatoms. The cell wall of a diatom is composed largely of silica rather than cellulose. Some diatoms move in a slow, gliding way as cytoplasm glides through slits in the cell wall to propel the organism.

The Pyrrhophyta are single-celled, algae-like organisms, the most important of which are the dinoflagellates (table 3.1). In most species of dinoflagellates, the cell wall is formed of armor-like plates of cellulose. Dinoflagellates are motile, having two flagella, one encircling the organism in a transverse groove, and the other projecting to the posterior.

Protozoa are also protists. They are small (2 μm—100 μm), unicellular eukaryotic organisms that lack a cell wall. Movement of protozoa is due to flagella, cilia, or pseudopodia of various sorts. In feeding upon other organisms or organic particles, they utilize simple diffusion, pinocytosis (active transport), or phagocytosis. Although most protozoa reproduce asexually, some species may also reproduce sexually during a portion of their life cycles. Most protozoa are harmless, although some are of immense clinical concern because they are parasitic and may cause human disease, including African sleeping sickness and malaria.

Table 4.1. Some Representatives of the Kingdom Protista

Phyla and representative kinds	Characteristics
Unicellular Protists	
Chrysophyta — diatoms and golden algae	Diatom cell walls of silica, with two halves; plastids often golden
Pyrrhophyta — dinoflagellates	Two flagella in grooves of wall; brownish plastids
Rhizopoda — amoebas	Cytoskeleton of microtubules and microfilaments; amoeboid locomotion
Apicomplexa — sporozoa, Plasmodium	Lack locomotor capabilities and contractile vacuoles; mostly parasitic
Euglenophyta — euglenoids	Green flagellates lacking typical cell walls
Ciliophora — ciliates, Paramecium	Use cilia to move and feed
Algae	
Chlorophyta — green algae	Unicellular, colonial, and multicellular forms, mostly freshwater; reproduce asexually and sexually; gametes often biflagellated, with cup–shaped chloroplasts
Phaeophyta — brown algae, giant kelp	Multicellular, mostly marine in the intertidal zone; alternation of generations common
Rhodophyta — red algae	Multicellular, mostly marine; sexual reproduction but with no flagellated cells; alternation of generations common
Protists Resembling Fungi	
Myxomycota — plasmodial slime molds	Multinucleated continuum of cytoplasm without internal membranes; amoeboid plasmodium during feeding stage; produce asexual, fruiting bodies
Acrasiomycota — cellular slime molds	Solitary cells during feeding stage; aggregate of cells when food is scarce; produce asexual, fruiting bodies
Oomycota — water molds, white rusts, downy mildews	Decomposers or parasitic forms; walls of cellulose; dispersal by spores or flagellated zoospores

Chrysophyta — diatoms and golden algae

Photo courtesy of Samuel R. Rushforth

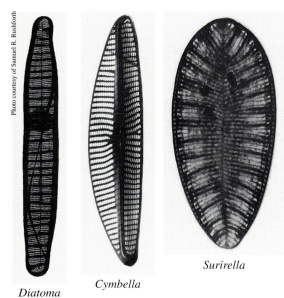

Diatoma

Cymbella

Surirella

Figure 4.1 Electron micrographs of several types of diatoms.

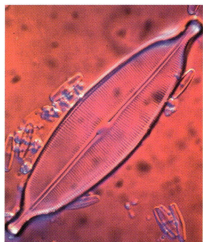

Eunotia

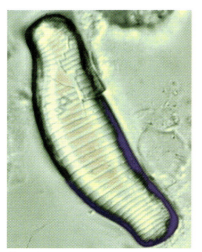

Navicula

Figure 4.2 Examples of common fresh-water diatoms. *Eunotia* species are often found in acidic waters common in forest ponds or bogs. *Navicula* is a large and widely distributed genus with species found in both marine and fresh-water habitats and on wet soil. *Cyclotella* and *Stephanodiscus* are centric (round in face) diatoms, common in lake plankton.

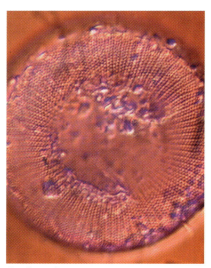

Stephanodiscus

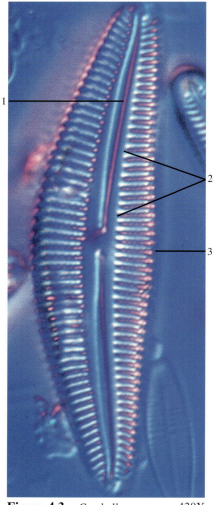

Figure 4.3 *Cymbella*, a common diatom.   430X

1. Raphe   2. Striae   3. Valve

Photo courtesy of James V. Allen

Figure 4.4 A scanning electron micrograph of the diatom *Achnanthes flexella*.   700X

1. Raphe   2. Striae

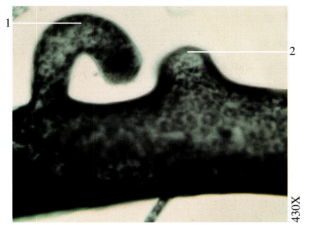

Figure 4.5 A filament with an immature gametagia of the "water felt," *Vaucheria*. *Vaucheria* is a chrysophyte that is widespread in fresh-water and marine habitats. It is also found in the mud of brackish areas that periodically become submerged and then exposed to air.

1. Antheridium   2. Oogonium

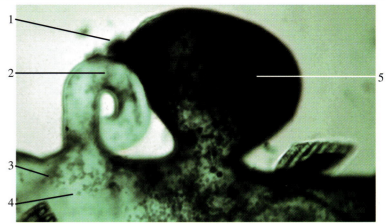

Figure 4.6   *Vaucheria*, with mature gametangia.   430X

1. Fertilization pore   3. Chloroplast   5. Oogonium
2. Antheridium   4. Coenocytic filament

Pyrrhophyta — dinoflagellates

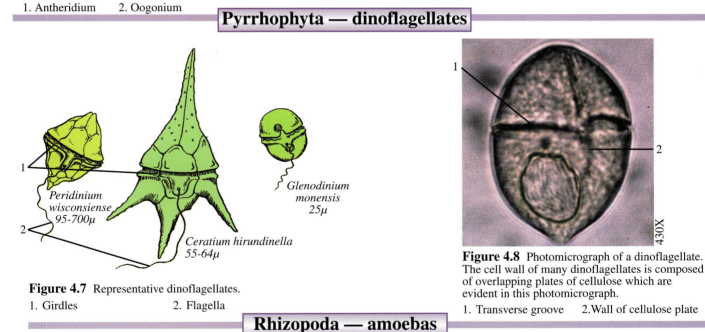

Peridinium wisconsiense 95-700µ

Ceratium hirundinella 55-64µ

Glenodinium monensis 25µ

Figure 4.7 Representative dinoflagellates.

1. Girdles   2. Flagella

Figure 4.8 Photomicrograph of a dinoflagellate. The cell wall of many dinoflagellates is composed of overlapping plates of cellulose which are evident in this photomicrograph.

1. Transverse groove   2. Wall of cellulose plate

Rhizopoda — amoebas

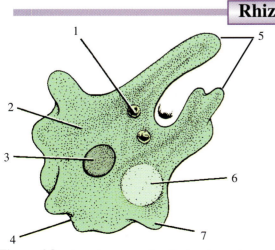

Figure 4.9 *Amoeba proteus* is a freshwater protozoan that moves by forming cytoplasmic extensions called pseudopodia.

1. Food vacuole   4. Cell membrane   6. Contractile
2. Endoplasm   5. Pseudopodia   vacuole
3. Nucleus   7. Ectoplasm

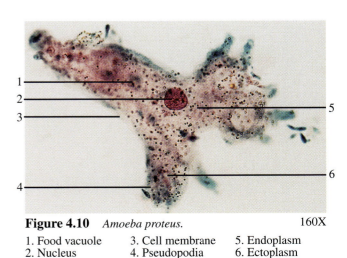

Figure 4.10   *Amoeba proteus.*   160X

1. Food vacuole   3. Cell membrane   5. Endoplasm
2. Nucleus   4. Pseudopodia   6. Ectoplasm

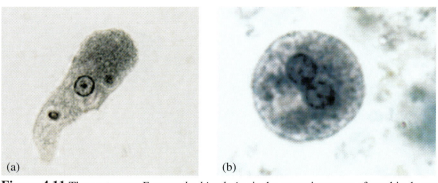

(a)   (b)

Figure 4.11 The protozoan *Entamoeba histolytica* is the causative agent of amebic dysentery, a disease most common in areas with poor sanitation. (a) A trophozoite and (b) a cyst.

Apicomplexa—sporozoans and *Plasmodium*

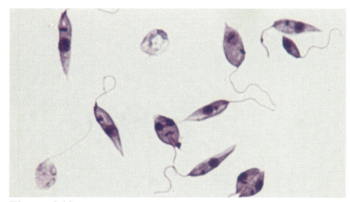

Figure 4.12 The protozoan *Leishmania donovani* is the causative agent of visceral leishmaniasis, or kala-azar disease. The sandfly is the infectious host of this disease.

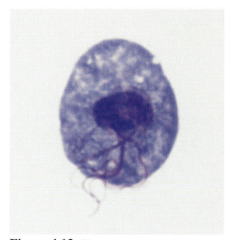

Figure 4.13 The protozoan *Trichomonas vaginalis* is the causative agent of trichomoniasis. Trichomoniasis is an inflammation of the genitourinary mucosal surfaces – the urethra, vulva, vagina, and cervix in females and the urethra, prostate, and seminal vesicles in males.

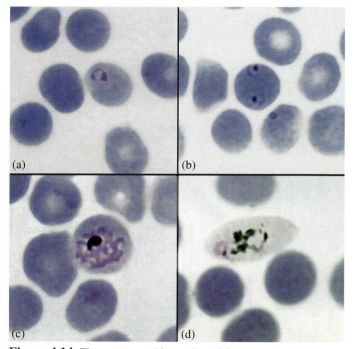

(a)   (b)

(c)   (d)

Figure 4.14 The protozoan *Plasmodium falciparum* causes malaria, which is transmitted by the female *Anopheles* mosquito. (a) The ring stage in a red blood cell, (b) a double infection, (c) a developing schizont, and (d) a gametocyte.

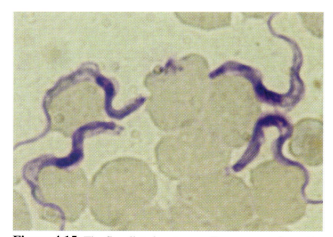

Figure 4.15 The flagellated protozoan *Trypanosoma brucei* is the causative agent of African trypanosomiasis, or African sleeping sickness. The tsetse fly is the infectious host of this disease.

Euglenophyta — euglenoids

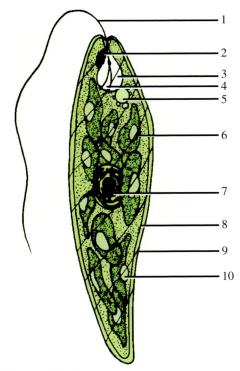

Figure 4.16 *Euglena* is a green flagellate that contains chloroplasts. They are freshwater organisms that have a flexible pellicle rather than a rigid cell wall.

1. Flagellum
2. Photoreceptor
3. Reservoir
4. Basal body
5. Contractile vacuole
6. Chloroplast
7. Nucleus
8. Pellicle
9. Cell membrane
10. Paramylon granule

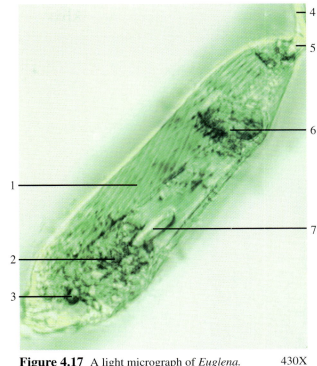

Figure 4.17 A light micrograph of *Euglena*.   430X

1. Striated pellicle
2. Chloroplast
3. Paramylum granule
4. Flagellum
5. Reservoir
6. Nucleus
7. Paramylum granule

Figure 4.18 *Euglena.*

Ciliophora—ciliates and *Paramecium*

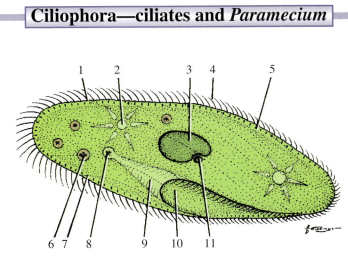

Figure 4.19 *Paramecium caudatum* is a ciliated protozoan. The poisonous trichocysts of these unicellular organisms are used for defense and capturing prey.

1. Pellicle
2. Contractile vacuole
3. Macronucleus
4. Cilia
5. Trichocyst
6. Food vacuole
7. Anal pore
8. Forming food vacuole
9. Gullet
10. Oral cavity
11. Micronucleus

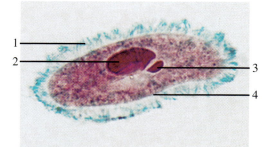

Figure 4.20 *Paramecium busaria* is   430X
a unicellular, slipper-shaped organism.
Paramecia are usually common in ponds
containing decaying organic matter.

1. Cilia   3. Micronucleus
2. Macronucleus   4. Pellicle

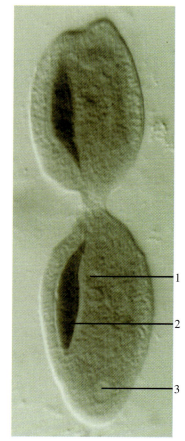

Figure 4.22   430X
Paramecium in fission.

1. Micronucleus
2. Macronucleus
3. Contractile vacuole

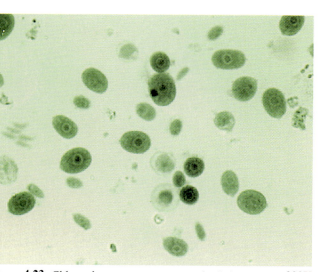

Figure 4.21 *Balantidium coli* is the causative agent of balantidiasis. Cysts in
sewage-contaminated water are the infective form.

Chlorophyta — green algae

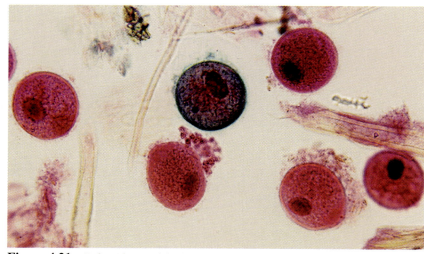

Figure 4.23 *Chlamydomonas*, a common unicellular   800X
green alga.

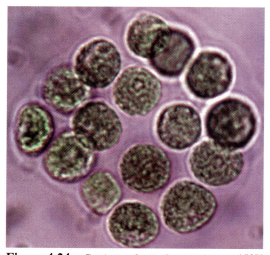

Figure 4.24 *Gonium* colony. *Gonium* is   450X
a 16-celled flat colony of *Chlamydomonas*-like cells.

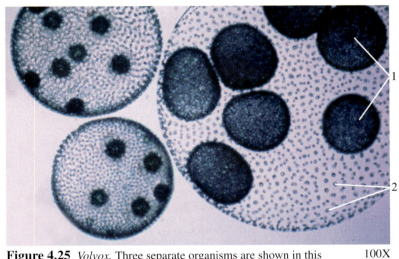

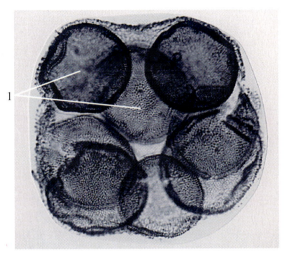

Figure 4.25 *Volvox.* Three separate organisms are shown in this   100X
photomicrograph, each containing daughter colonies.

1. Daughter colonies   2. Vegetative Cells

Figure 4.26 *Volvox,* a single organism with   100X
several large daughter colonies.

1. Daughter colonies

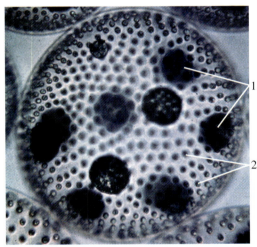

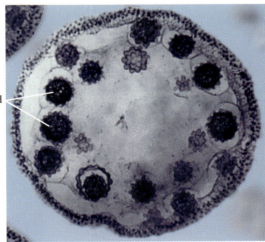

Figure 4.27 *Volvox,* a single mature   100X
specimen with several eggs and zygotes.

1. Zygotes   2. Vegetative cells

Figure 4.28 *Volvox,* a single mature   100X
organism with zygospores.

1. Zygospores

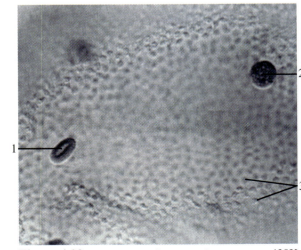

Figure 4.29 *Volvox,* a motile green alga.   430X
This photomicrograph is a highly magnified view of a
single organism.

1. Sperm packet   2. Egg   3. Vegetative cells

Figure 4.30 *Ulothrix* is an unbranched, filamentous green alga.   100X

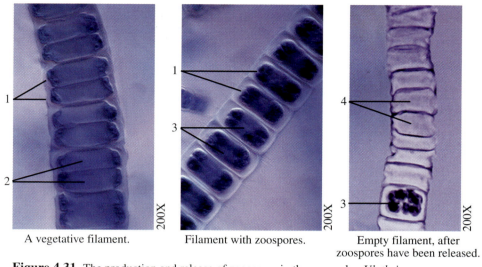

A vegetative filament.   Filament with zoospores.   Empty filament, after
                                                                   zoospores have been released.

Figure 4.31 The production and release of zoospores in the green alga *Ulothrix.*

1. Filament   2. Chloroplasts   3. Zoospores   4. Empty cells

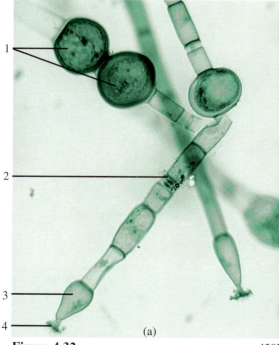

(a)

Figure 4.32   430X

(a) *Oedogonium,* a
filamentous,
unbranched, green
alga. (b) A closeup
of an oogonium.

1. Oogonia
2. Antheridium
3. Basal cell
4. Holdfast

(b)

1000X

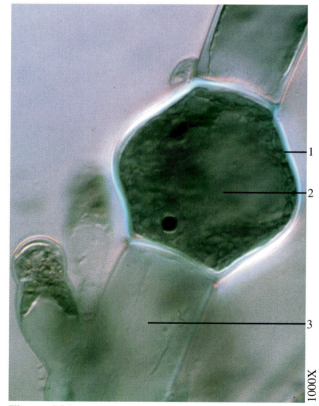

Figure 4.33 The oogonium of the unbranched, green alga,
Oedogonium.

1. Oogonium   2. Egg   3. Vegetative cell

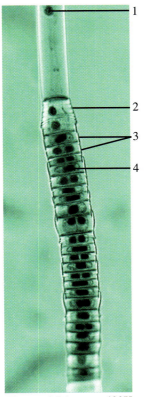

Figure 4.34   430X
A filament of the
unbranched green alga,
Oedogonium.

1. Cell nucleus
2. Annular scars from
   cell division
3. Antheridia
4. Sperm

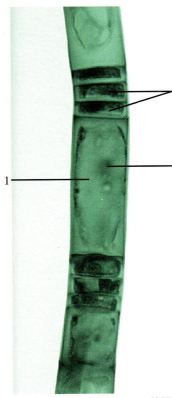

Figure 4.35   600X
The green alga, *Oedogonium,*
showing antheridia between
vegetative cells.

1. Vegetative cell
2. Antheridia
3. Sperm

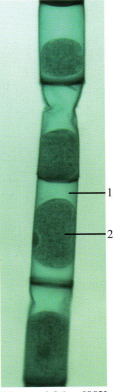

Figure 4.36   600X
The zoosporangium of
the unbranched green
alga, *Oedogonium.*

1. Zoosporangium
2. Zoospore

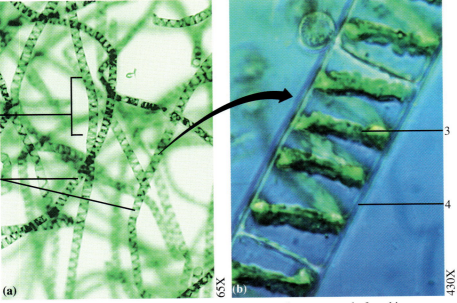

Figure 4.37   Species of *Spirogyra* are filamentous green algae commonly found in green
masses on the surfaces of ponds and streams. Their chloroplasts are arranged as a spiral
within the cell. (a) Several cells comprise a filament. (b) A magnified view of a single
filament composed of several cells.

1. Single cell   2. Filaments   3. Chloroplast   4. Cell wall

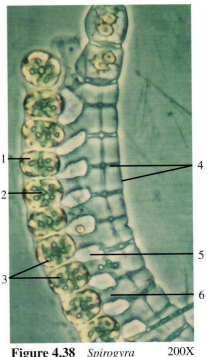

Figure 4.38   *Spirogyra*   200X
undergoing conjugation.

1. Zygote   4. Cell wall
   (zygospore)   5. Male gamete
2. Chloroplast   6. Conjugation
3. Pyrenoids   tube

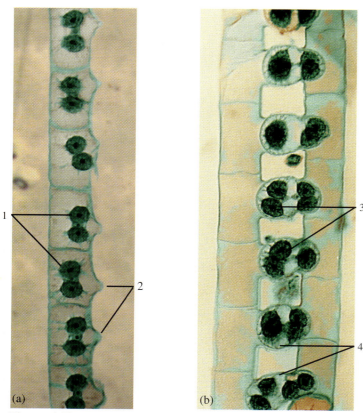

Figure 4.39　*Zygnema* undergoing conjugation.　　　100X
(a) The filament is just forming conjugation tubes; (b) and two conjugated filaments. Zygotes in this species are produced in the conjugation tubes.

1. Gametes
2. Forming conjugation tubes
3. Forming zygotes
4. Conjugation tubes

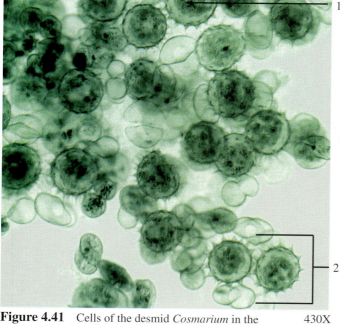

Figure 4.41　Cells of the desmid *Cosmarium* in the　　430X
process of conjugation.

1. Zygote produced from conjugation
2. Parent cells that have produced gametes

Figure 4.40　　　　　　　　200X
Representative desmids. Desmids are unicellular, freshwater chlorophyta.

1. *Closterium*

Figure 4.42　Sea lettuce, *Ulva,* lives as a flat membranous form in marine environments.

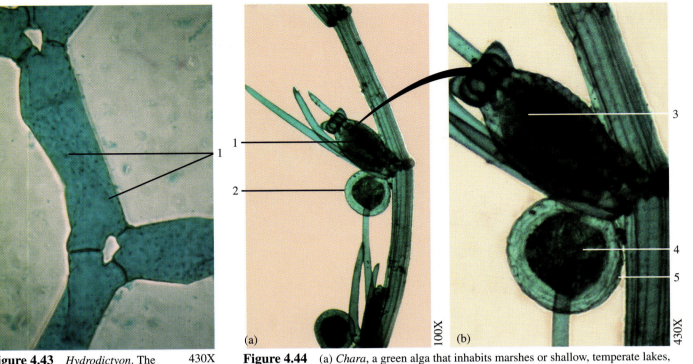

Figure 4.43   *Hydrodictyon*. The large, multinucleated cells form net-shaped colonies.   430X

1. Nuclei

Figure 4.44   (a) *Chara*, a green alga that inhabits marshes or shallow, temperate lakes, showing characteristic gametangia. (b) A magnified view of the gametangia.

1. Oogonium   3. Oogonium   5. Antheridium
2. Antheridium   4. Egg

Phaeophyta — brown algae

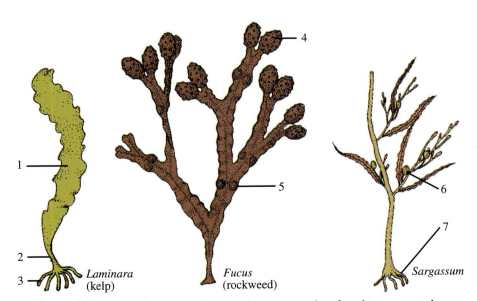

Laminara (kelp)   *Fucus* (rockweed)   *Sargassum*

Figure 4.45   Brown algae are marine organisms commonly referred to as seaweeds. They range in size from small filamentous organisms a few millimeters in length to the giant kelp between 50 and 100 meters long.

1. Blade   4. Receptacle   6. Air bladder
2. Stipe   5. Air bladder   7. Holdfast
3. Holdfast

Figure 4.46   "Sea palm," *Postelsia palmaeformis*, a common brown alga found on the west coast of North America.

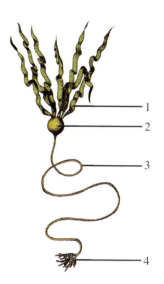

Figure 4.47 The brown alga *Nereocystis* has a long stipe and photosynthetic laminae attached to a large float. The holdfast anchors the alga to the ocean floor. This and other brown algae can grow to lengths of several meters.

1. Lamina   3. Stipe
2. Float   4. Holdfast

Figure 4.49 *Sargassum*, a brown alga.

1. Float (air-filled bladder)
2. Blade
3. Stipe

Figure 4.48 An algal hummock, formed by detached brown algae washing ashore and becoming entangled.

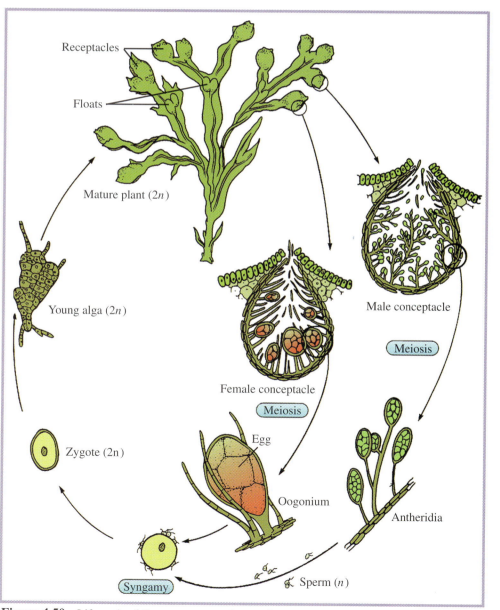

Figure 4.50 Life cycle of *Fucus*, a common brown alga.

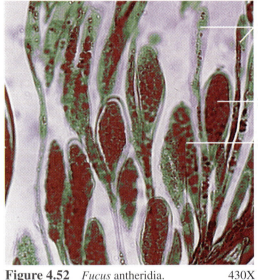

Figure 4.51   (a) *Fucus*, a brown alga, commonly called rockweed. (b) An enlargement of a blade supporting the receptacles.

1. Blade
2. Receptacle
3. Stipe

4. Conceptacles (light-colored spots) are chambers imbedded in the receptacles
5. Blade

Figure 4.52   *Fucus* antheridia.   430X
1. Paraphyses
2. Antheridium
3. Sperm within antheridium

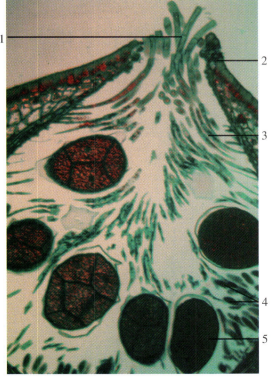

Figure 4.53   *Fucus*, conceptacle   200X
containing both antheridia and oogonia.

1. Ostiole
2. Surface of receptacle
3. Paraphyses (sterile hairs)

4. Antheridia
5. Oogonium

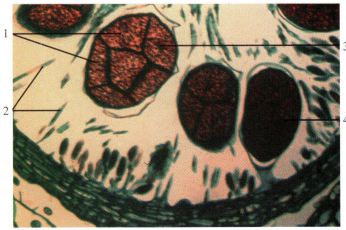

Figure 4.54   *Fucus*, closeup of female conceptacle.   200X

1. Eggs
2. Paraphyses

3. Nucleus of egg
4. Oogonium

Rhodophyta — red algae

Porphyra   *Rhodomenia*   *Polysiphonia.*

Figure 4.55 Examples of common marine red algae.

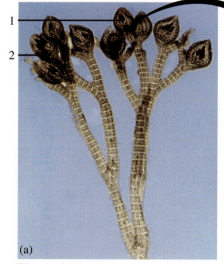

(a)

(b)

Figure 4.56 The red alga,   100X
Polysiphonia, has alternation of three
generations. (a) Female gametophyte
with attached carposporophyte
generation. (b) A closeup of the
cystocarps.

1. Pericarp
2. Carposporophyte producing
   carpospores
3. Cystocarp
4. Carpospores

Figure 4.57 *Polysiphonia*, 100X
tetrasporophyte.

1. Tetrasporophyte (2n)
2. Tetraspores (n)

Figure 4.58 *Polysiphonia*,   40X
male gametophyte. Male reproductive
structures, known as spermatangia,
produce non-motile spermatia.
1. Spermatangia

Myxomycota—plasmodial slime molds

Figure 4.59   Slime mold sporangia vary considerably in size and shape. (a) a species of *Fuligo*; (b) a species of *Lycoperdon*; and (c) a species of *Lycogala.*

Spores

Sporangium

Figure 4.60   Diagrams of the slime mold, *Hemitrichia*, showing the release of spores from the sporangium.

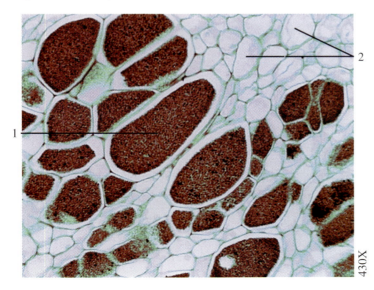

430X

Figure 4.61   *Plasmodiophora brassicae*, a myxomycete responsible for clubroot in cabbage. Depicted is a cross section through a cabbage root showing spores of *Plasmodiophora* in the cabbage root cells.

1. Spores   2. Cabbage root cells.

Figure 4.62   Sporangia of the slime mold *Comatricha typhoides*.

Figure 4.63   Slime mold, *Physarum*, growing on oats and an agar medium.

Oomycota — water molds, white rusts, and downy mildews

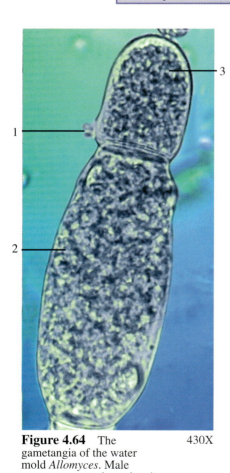

Figure 4.64   The 430X
gametangia of the water
mold *Allomyces*. Male
gametes escape through exit pores.

1. Exit pore
2. Female gametangium
3. Male gametangium

Figure 4.65   A water mold,   430X
Saprolegnia, showing a young
oogonium before eggs have
been formed.

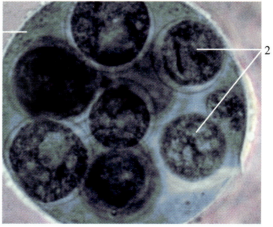

Figure 4.66   A mature oogonium of the   430X
water mold *Saprolegnia*.

1. Oogonium   2. Eggs

Kingdom Fungi

5

About 250,000 species of fungi exist. All fungi are heterotrophs because they absorb nutrients through their cell walls and cell membranes. The kingdom Fungi includes the typical conjugation fungi, yeasts, mushrooms, toadstools, rusts, and lichens. Most are saprobes, absorbing nutrients from dead organic material, while a few are parasitic, absorbing nutrients from living hosts.

Except for the unicellular yeasts, fungi consist of elongated filaments called *hyphae*. Hyphae begin as tubular extensions of spores that branch as they grow to form a network of hyphae called a *mycelium*. Even the body of a mushroom consists of a mass of tightly packed hyphae attached to an underground mycelium. Fungi are nonmotile and reproduce by means of spores, which are produced sexually or asexually.

Fungi help decompose organic material, helping to recycle the inorganic nutrients essential for plant growth.

Many species of fungi are commercially important. Some are used as food, such as mushrooms; or in the production of foods, such as the use of yeasts in making bread, cheese, beer, and wine. Other species are important in medicine, for example, in the production of antibiotic penicillin. Many other species of fungi are of medical and economic concern because they cause plant and animal diseases and destroy crops and stored goods.

Table 5.1 Some Representatives of the Kingdom Fungi

Divisions and representative kinds	Characteristics
Zygomycota — conjugation fungi	Hyphae lack cross walls between nuclei
Ascomycota — yeasts, molds, morels, truffles	Septate hyphae, reproductive structures contain 8 ascospores within asci; asexual reproduction by budding or conidia
Basidiomycota — mushrooms, toadstools, rusts, smuts	Septate hyphae; 4 spores produced externally on cells called basidia contained in basidiocarp (basidioma)
Lichens — not a division, but rather a symbiotic association of an alga and a fungus	Algal component (usually a green alga) provides food from photosynthesis; fungal component (usually an ascomycete) may provide anchorage, water retention, and/or nutrient absorbance

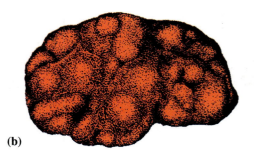

Spores

(a)

(b)

Figure 5.1 Diagrams of two commercially important ascomycetes. (a) An ascocarp of *Morchella esculenta,* the common morel prized as a gourmet food; (b) an ascocarp of the truffle *Tuber.* Truffles develop their ascocarps underground, where they are difficult to find and harvest.

Zygomycota — conjugation fungi

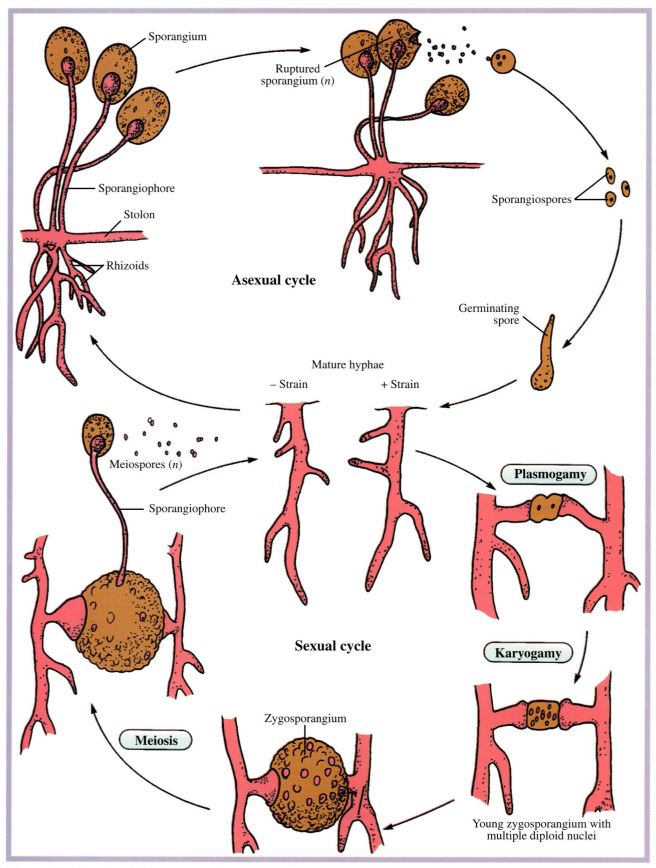

Figure 5.2 Life cycle of *Rhizopus*, the common bread mold.

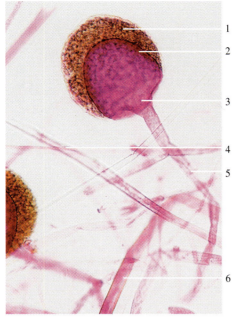

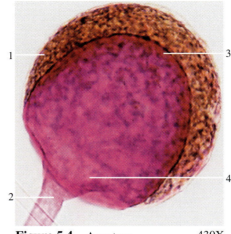

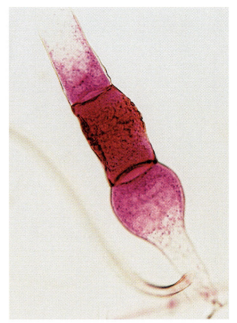

Figure 5.4   A mature   430X
sporangium in the asexual reproductive
cycle of the bread mold *Rhizopus.*

1. Sporangium   3. Spores
2. Sporangiophore   4. Columella

Figure 5.3   A whole mount of   100X
the bread mold *Rhizopus.*

1. Sporangium   4. Rhizoid
2. Spores   5. Sporangiophore
3. Columella   6. Stolon (hyphae)

Figure 5.5   An immature *Rhizopus*   264X
zygosporangium (zygospore).

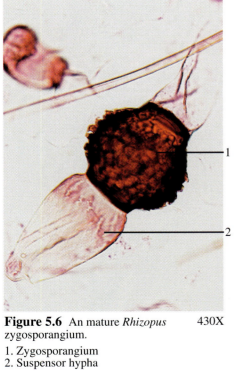

Figure 5.6 An mature *Rhizopus*   430X
zygosporangium.

1. Zygosporangium
2. Suspensor hypha

Ascomycota – yeasts, molds, morels, and truffles

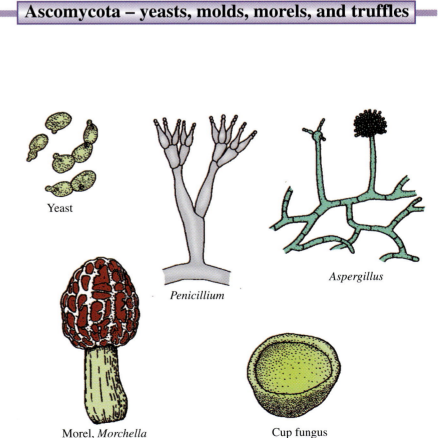

Figure 5.7 Various ascomycetes.

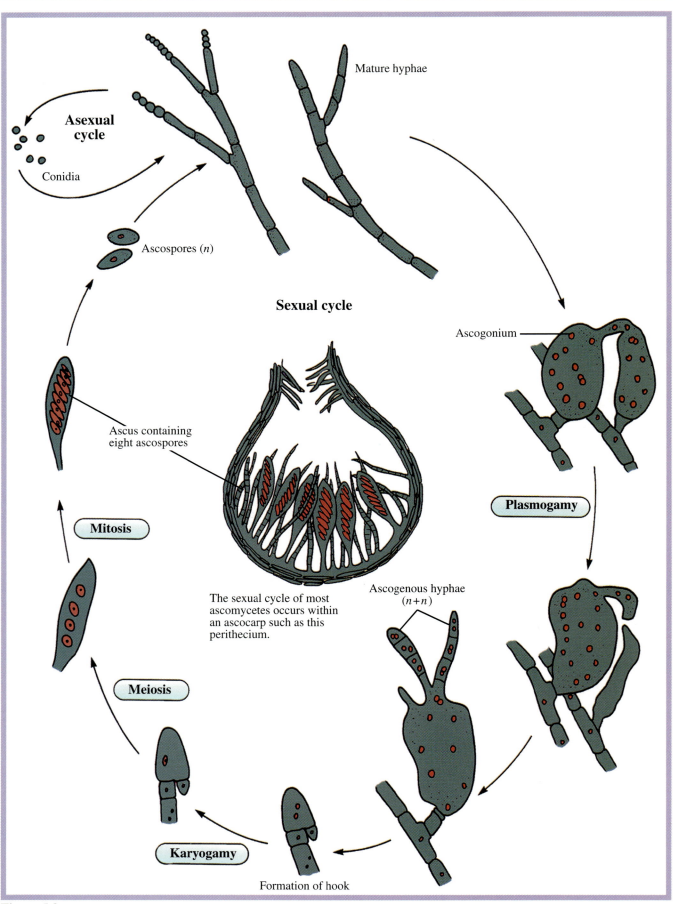

Figure 5.8 Life cycle of an ascomycete.

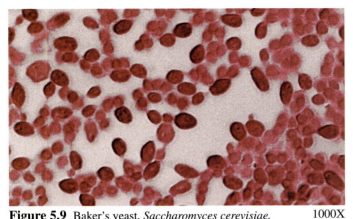

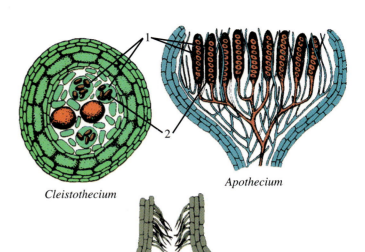

Cleistothecium   *Apothecium*

Figure 5.9 Baker's yeast, *Saccharomyces cerevisiae.*
The ascospores of this unicellular ascomycete are
characteristically spheroidal or ellipsoidal in shape.   1000X

Perithecium

Figure 5.10 Examples of various types of ascocarps, or fruiting
bodies, of ascomycetes.

1. Asci   2. Ascospores

Morchella   *Helvella*

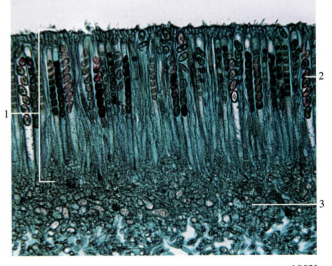

Peziza repanda   *Monolina fructicola*

Figure 5.11 Fruiting bodies (ascocarps) of common ascomycetes.
Morchella is a common edible morel, *Helvella* is sometimes known as a
saddle fungus since the fruiting body is thought by some to resemble a
saddle, *Peziza repanda* is a common woodland cup fungus, and
Monolina fructicola is an important plant pathogen causing brown rot of
fruit.

1. Pit of fruit

Figure 5.12 A section through the hymenial layer   100X
of the apothecium of *Peziza,* showing asci with ascospores.

1. Hymenial layer
2. Ascus with ascospores
3. Ascocarp mycelium

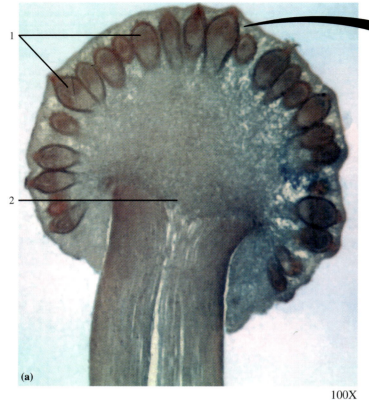

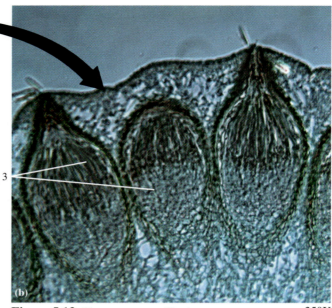

(a)

100X

Figure 5.13 The ascomycete *Claviceps purpurea*.   250X
(a) A longitudinal section through the ascocarps.
(b) An enlargement of the perithecia. This fungus causes serious plant diseases and is toxic to humans.

1. Perithecia
2. Stroma
3. Perithecia containing asci with ascospores

Photographs courtesy of James V. Allen

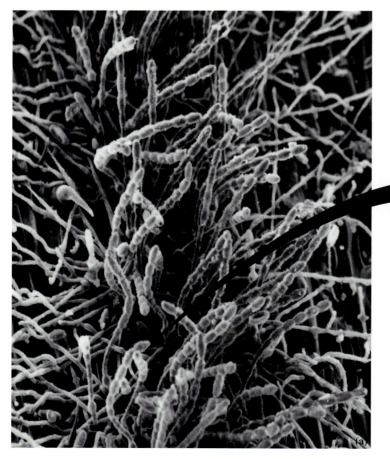

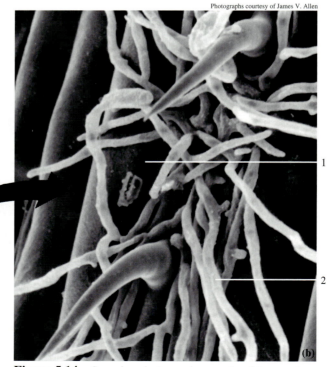

Figure 5.14 Scanning electron micrographs of the powdery mildew, *Erysiphe graminis,* on the surface of wheat. As the mycelium develops, it produces spores (conidia) that give a powdery appearance to the wheat.

1. Wheat host
2. Mycelium

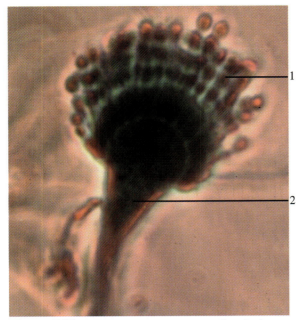

Figure 5.15 The common mold *Aspergillus.* A closeup of a conidiophore with chains of asexual conidia (spores) at the end.

1. Conidia
2. Conidiophore

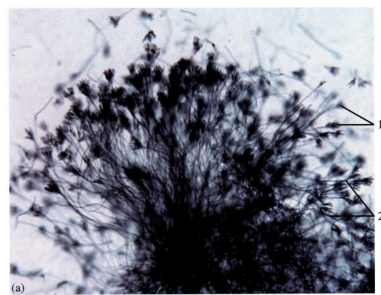

(a)   100X

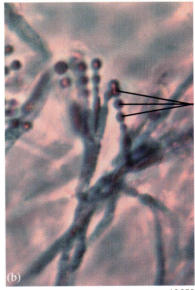

(b)   430X

Figure 5.16 The fungus *Penicillium* causes economic damage as a mold but is also the source of important antibiotics. (a) A colony of *Penicillium* and (b) a closeup of a conidiophore with chains of asexual spores (conidia) at the end.

1. Conidia
2. Conidiophores
3. Chains of asexual spores

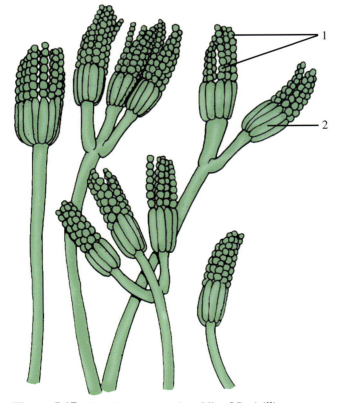

Figure 5.17 Conidiophores and conidia of *Penicillium.*

1. Conidia
2. Conidiophore

Busidiomycota — mushrooms, toadstools, rusts, and smuts

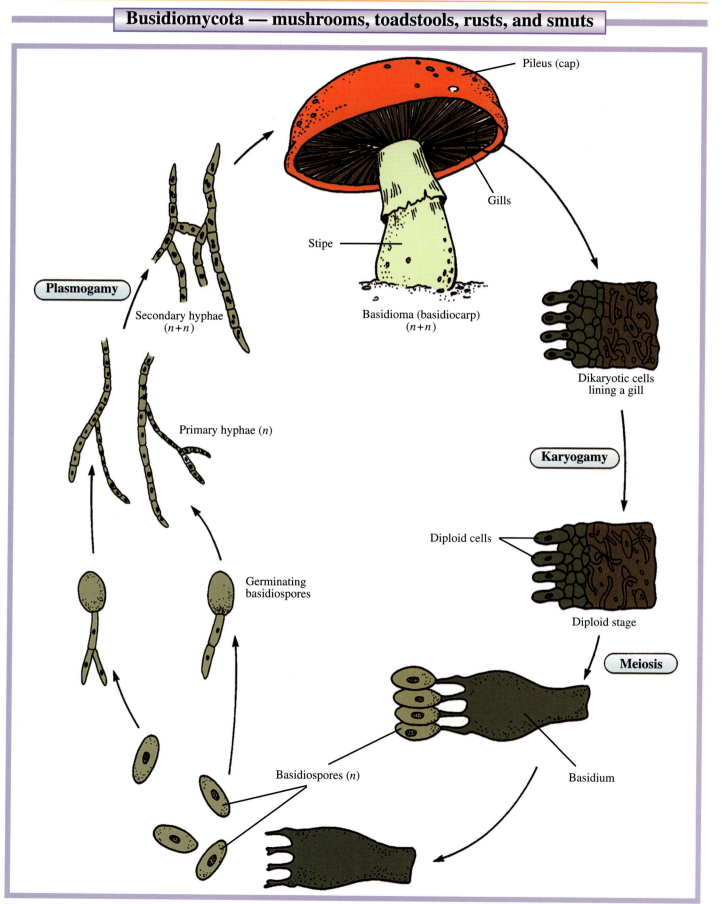

Pileus (cap)

Gills

Stipe

Basidioma (basidiocarp)
($n+n$)

Dikaryotic cells
lining a gill

Plasmogamy

Secondary hyphae
($n+n$)

Karyogamy

Primary hyphae (n)

Diploid cells

Diploid stage

Germinating
basidiospores

Meiosis

Basidiospores (n)

Basidium

Figure 5.18 Life cycle of a mushroom.

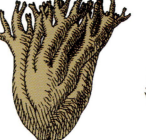

Shelf fungi
Ganoderma applanatum

Coral fungi
family *Clavariaceae*

Puffball
Lycoperdon ericetorum

Earthstars
Geastrum saccutum

Figure 5.19   Drawings of various basidiomycetes.

Agaricus rodmani

Cortinarius

Amanita pantherina

Boletus

Figure 5.20   Fruiting bodies (basidiocarps) of common mushrooms. *Agaricus rodmani* is a prized edible mushroom, the genus *Cortinarius* contains more mushrooms than any other in North America, *Amanita pantherina* is a deadly toxic mushroom, and *Boletus* is a mushroom with pores on the undersurface of the pileus (cap) rather than on gills.

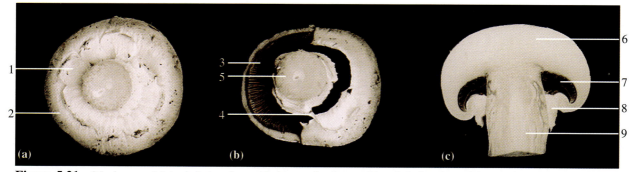

Figure 5.21   Mushroom. (a) An inferior view with the annulus intact, (b) an inferior view with a portion of the annulus removed to show the gills, and (c) a longitudinal section.

1. Veil
2. Pileus (cap)
3. Gills

4. Stipe (stalk)
5. Annulus
6. Pileus (cap)

7. Gills
8. Veil
9. Stipe (stalk)

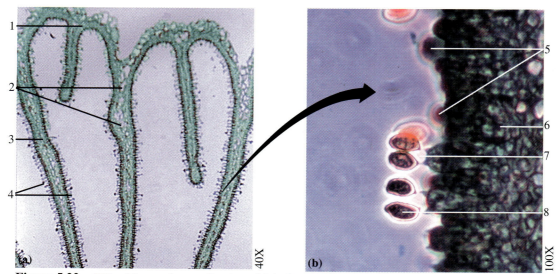

Figure 5.22   Gills of the mushroom *Coprinus*. (a) A closeup of several gills and (b) a closeup of a portion of a single gill.

1. Pileus (cap) comprised of gills
2. Hyphae comprising the gills
3. Gill

4. Basidiospores
5. Basidia

6. Gill (comprised of hyphae)
7. Sterigma
8. Basidiospore

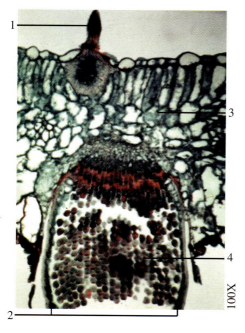

Figure 5.23
Wheat rust, *Pycnium graminis,* pycnium and aecia on barberry leaf.
1. Pycnidium
2. Aecium
3. Barberry leaf
4. Aeciospores

Figure 5.22 A photograph of a corn plant infected by the smut *Ustilago maydis.*

1. Corn stalk
2. A corn ear destroyed by the fungus

Figure 5.25 A photograph of a smut-infected brome grass. The grains have been destroyed by the fungus.

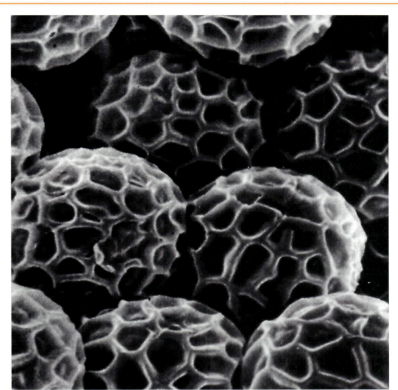

Figure 5.26 A scanning electron micrograph of teliospores of a wheat smut fungus.

Lichens

Crustose lichen

Figure 5.27 Although highly diverse in structure and appearance, the many kinds of lichens are grouped into three broad categories.

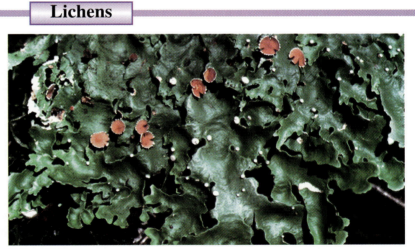

Foliose lichen

Fruticose lichen

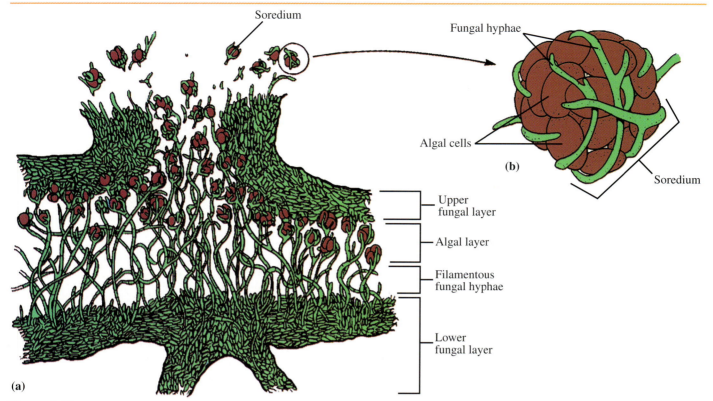

(a)

Figure 5.28 Many lichens reproduce by producing soredia, which are small bodies containing both algal and fungal cells. (a) A lichen thallus and (b) a soredium.

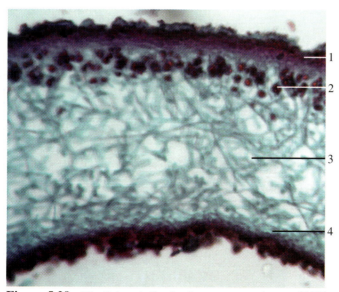

Figure 5.29 The lichen thallus is often constructed   100X
of distinct algal and fungal layers. This cross section of
a lichen thallus clearly shows these layers.

1. Upper fungal layer   3. Filamentous fungal hyphae
2. Algal layer   4. Lower fungal layer

Figure 5.30   A crustose lichen of the bark of a tree.

1. Lichen

Kingdom Plantae

6

Plants are photosynthetic, multicellular eukaryotes. *Cellulose* in their cell walls provides protection and rigidity, while the *stomata* and *cuticle* of stems and leaves regulate gas exchange. Mitosis and meiosis are characteristic of all plants. Jacketed sex organs, called *gametangia*, protect the gametes and embryos from desiccation. Most plants have heteromorphic alternation of generations with distinctive haploid *gametophyte* and diploid *sporophyte* forms. Photosynthetic cells within plants contain *chloroplasts* with the pigments chlorophyll *a*, chlorophyll *b*, and a variety of carotenoids. Carbohydrates are produced by plants and stored in the form of starch.

Reproduction in seed plants is well adapted to a land existence. The conifers produce their seeds in protective *cones*, and the angiosperms produce their seeds in protective *fruits*. In the life cycle of a conifer, such as a pine, the mature *sporophyte* (tree) has female cones which produce *megaspores* that develop into the female gametophyte generation, and male cones which produce *microspores* that develop into the male gametophyte generation (mature pollen grains). Following fertilization, immature sporophyte generations are present in seeds located on the female cones. The female cone opens and the *seeds* (pine nuts) disperse to the ground and germinate if the conditions are right. Reproduction in angiosperms is similar to gymnosperms except that the angiosperm pollen and ovules are produced in flowers, rather than in cones and a fruit is formed.

Table 6.1 Some Representatives of the Kingdom Plantae

Divisions and representative kinds	Characteristics
Bryophyta — liverworts, hornworts, and mosses	Flattened and waxy leaflike cuticles that lack vascular tissue; rootlike rhizomes supporting rhizoids; homosporous
Psilotophyta — whisk ferns	True roots and leaves are absent, but vascular tissue present; rhizome and rhizoids present
Lycophyta — clubmosses and quillworts	Sporangia borne on sporophylls; homosporous (bisexual gametophyte); many are epiphytes
Sphenophyta — horsetails	Epidermis embedded with silica; tips of stems bear conelike structures containing sporangia; most homosporous
Pterophyta — ferns	Nonseed producing, vascular plants; fronds as leaves, underground rhizome as roots; homosporous
Cycadophyta — cycads	Heterosporous, pollen and seed cones borne of different plants; large pith
Ginkgophyta — ginkgo	Deciduous, fan-shaped leaves; seed-producing; heterosporous
Coniferophyta — conifers	Woody plants that produce their seeds in air spaces, and the stoma are sunken
Anthophyta — angiosperms (monocots and dicots)	Flowering plants that produce their seeds enclosed in fruit; heterosporous; most are free-living, some are saprophytic or parasitic

Table 6.2 Some Representatives of the Division Bryophyta

Classes and representative kinds	Characteristics
Hepaticae — liverworts	Flat or leafy gametophytes; single-celled rhizoids; simple sporophytes that lack stomata
Anthocerotae — hornworts	Flat, lobed gametophytes; more complex sporophytes with stomata
Musci — mosses	Leafy gametophytes, multicellular rhizoids; sporophytes with stomata

Bryophyta — liverworts, hornworts, and mosses

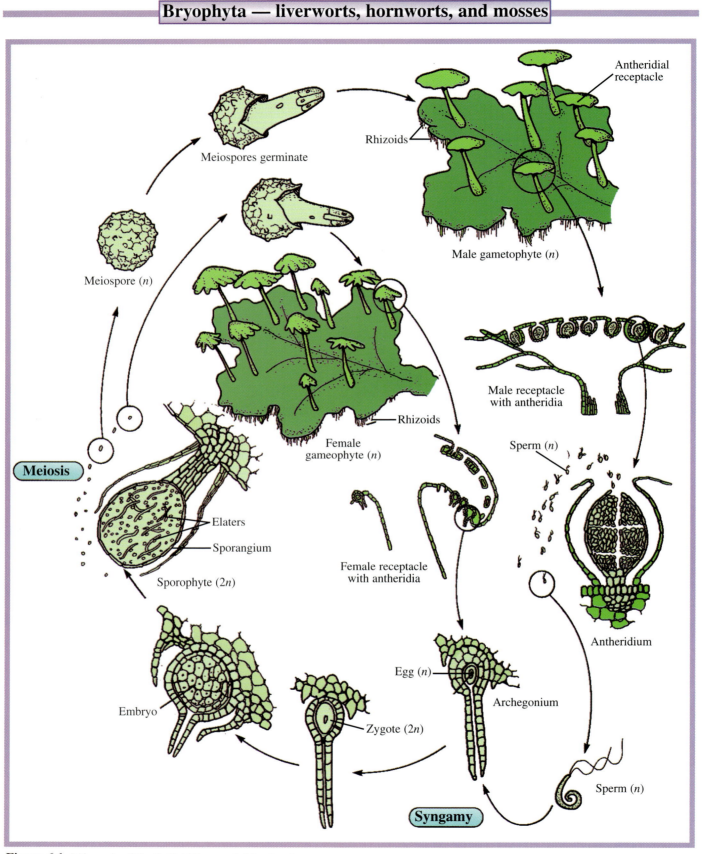

Meiospores germinate

Antheridial receptacle

Rhizoids

Male gametophyte (*n*)

Meiospore (*n*)

Male receptacle with antheridia

Sperm (*n*)

Meiosis

Female gameophyte (*n*)

Rhizoids

Elaters

Sporangium

Sporophyte (2*n*)

Female receptacle with antheridia

Antheridium

Egg (*n*)

Archegonium

Embryo

Zygote (2*n*)

Sperm (*n*)

Syngamy

Figure 6.1 Life cycle of the thalloid liverwort, *Marchantia.*

Figure 6.2 A photograph of several plants of the common liverwort *Marchantia.*

1. Antheridial receptacle   3. Gemmae cup (*n*)
2. Gametophyte thallus (*n*)

Figure 6.3 A dorsal view of a female gametophyte of the liverwort *Marchantia.*

1. Archegonial receptacles

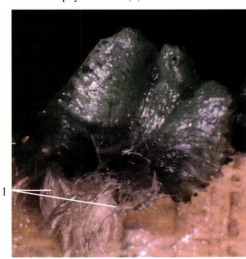

Figure 6.4 A ventral view of the liverwort *Marchantia,* showing numerous rhizoids.

1. Rhizoids

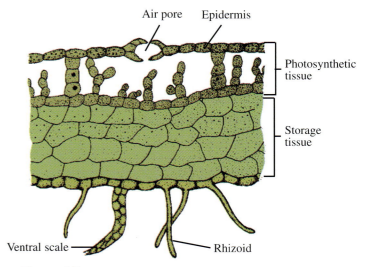

Figure 6.5 A diagram of the thallus of *Marchantia.*

Air pore   Epidermis

Photosynthetic tissue

Storage tissue

Ventral scale   Rhizoid

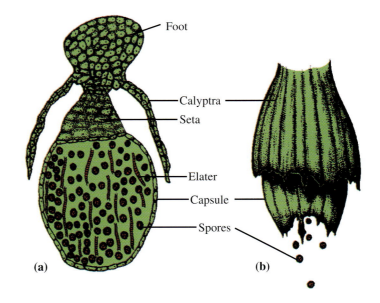

Figure 6.6 The sporophyte of *Marchantia,* (a) A longitudinal section and (b) a face view showing the shedding spores.

Foot

Calyptra

Seta

Elater

Capsule

Spores

(a)   (b)

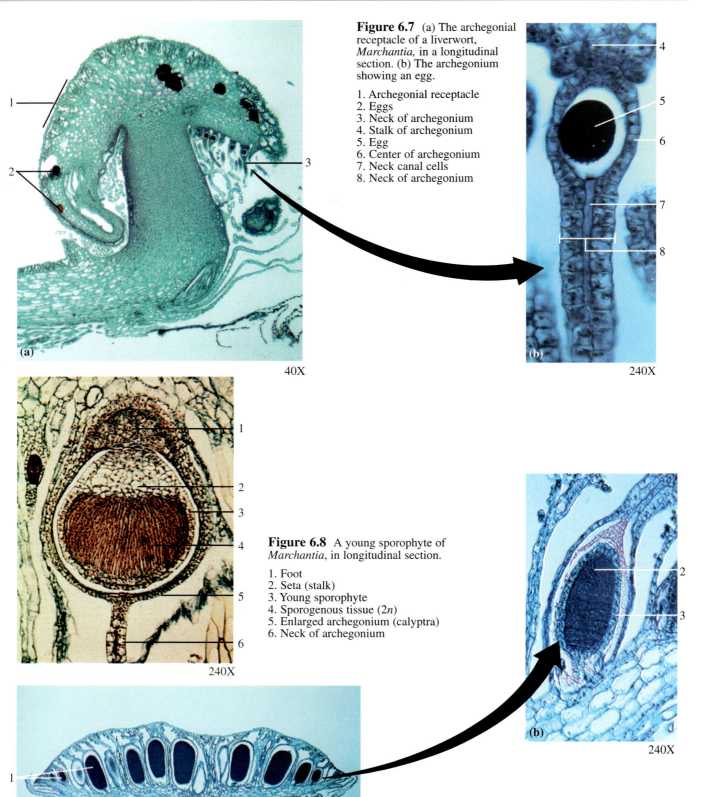

Figure 6.7 (a) The archegonial receptacle of a liverwort, *Marchantia*, in a longitudinal section. (b) The archegonium showing an egg.

1. Archegonial receptacle
2. Eggs
3. Neck of archegonium
4. Stalk of archegonium
5. Egg
6. Center of archegonium
7. Neck canal cells
8. Neck of archegonium

40X

240X

Figure 6.8 A young sporophyte of *Marchantia*, in longitudinal section.

1. Foot
2. Seta (stalk)
3. Young sporophyte
4. Sporogenous tissue (2*n*)
5. Enlarged archegonium (calyptra)
6. Neck of archegonium

240X

Figure 6.9 (a) The male receptacle with antheridia of a liverwort, *Marchantia*, in a longitudinal section. (b) The antheridial head showing a developing antheridium.

1. Antheridia
2. Spermatogenous tissue
3. Antheridium

240X

40X

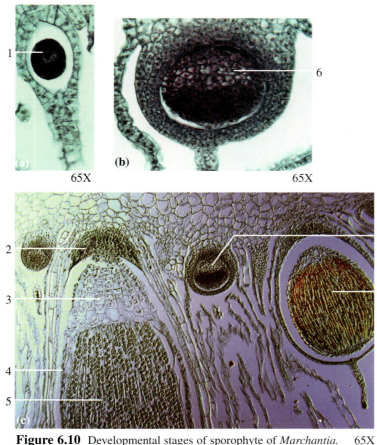

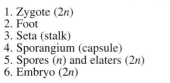

65X   65X

Figure 6.10 Developmental stages of sporophyte of *Marchantia*.   65X (a) Archegonium with zygote, (b) archegonium with embryos, and (c) immature and mature sporophytes.

1. Zygote (2*n*)
2. Foot
3. Seta (stalk)
4. Sporangium (capsule)
5. Spores (*n*) and elaters (2*n*)
6. Embryo (2*n*)
7. Immature sporophyte
8. Immature sporophyte with meiospore mother cells

Figure 6.11 A sporophyte of the leafy liverwort, *Parella*.

1. Sporophyte (2*n*)   3. Seta (stalk)
2. Capsule   4. Gametophyte (*n*)

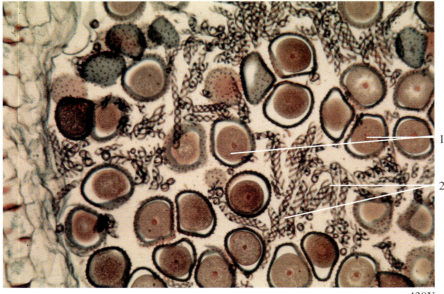

Figure 6.12 A cross section of the leafy liverwort, *Parella*.

1. Spores
2. Elaters

430X

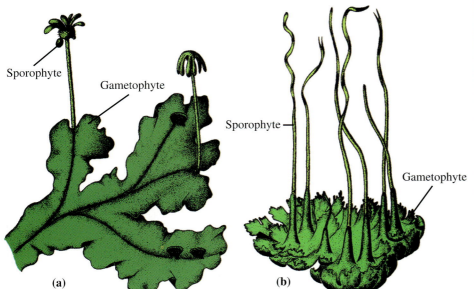

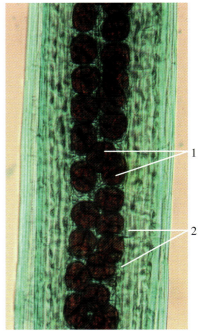

Sporophyte

Gametophyte

Sporophyte

Gametophyte

(a)   **(b)**

Figure 6.13 A comparison of the sporophytes and gametophytes of (a) the liverwort, *Marchantia,* and (b) the hornwort, *Anthoceros.*

Figure 6.14 A longitudinal   100X
section of the sporangium of a
sporophyte from the hornwort
Anthoceros.

1. Spores
2. Elater-like structures

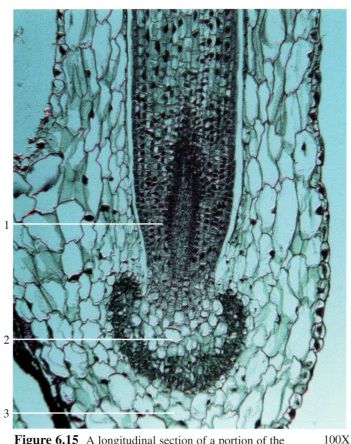

Figure 6.15 A longitudinal section of a portion of the   100X
sporophyte of the hornwort *Anthoceros.*

1. Meristematic region of sporophyte
2. Foot
3. Gametophyte

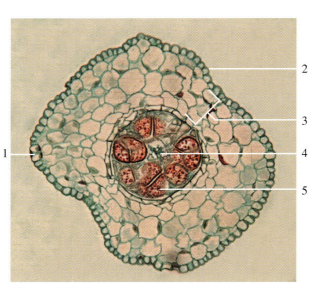

Figure 6.16 A transverse section through the   100X
capsule of a sporophyte of the hornwort *Anthoceros.*

1. Stoma   4. Columella
2. Epidermis   5. Spore
3. Photosynthetic tissue

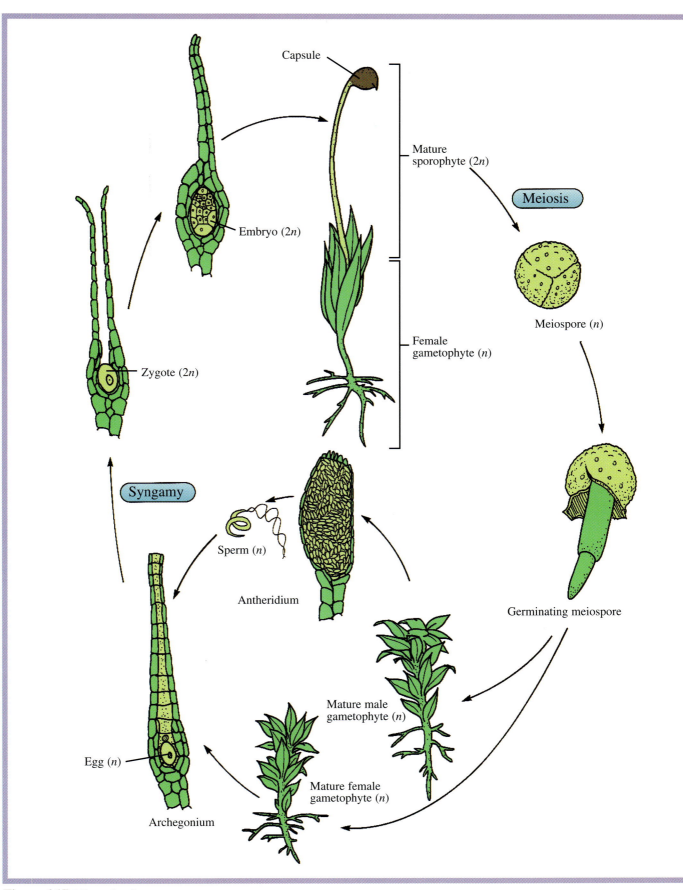

Figure 6.17 Life cycle of a moss.

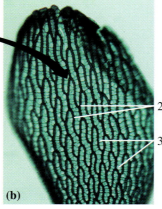

(b)

Figure 6.18 (a) A   40X
gametophyte of peat
moss, *Sphagnum.* (b) A
magnified view of a leaf,
showing the dead cell chambers

1. Leaves   3. Dead cells
2. Photosynthetic
   cells

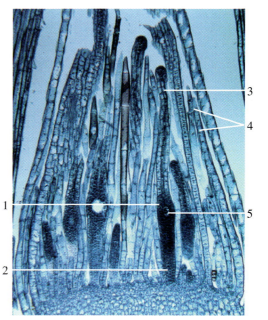

Figure 6.19 A longitudinal section of   180X
the archegonial head of the moss *Mnium.*
The paraphyses are non-reproductive filaments
that support the archegonia.

1. Venter   4. Paraphyses
2. Stalk   5. Egg
3. Neck

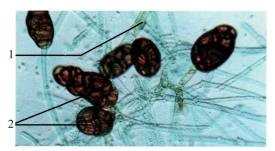

Figure 6.20 The gametophyta of a moss   430X
develops from buds along the protonema.
Several buds developing from a protonema are
illustrated here.

1. Protonema   2. Buds

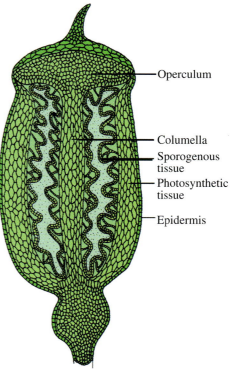

Operculum

Columella

Sporogenous
tissue

Photosynthetic
tissue

Epidermis

Figure 6.21 A longitudinal section of the antheridial   65X
head of the moss *Mnium.*

1. Spermatogenous tissue   4. Paraphyses (sterile filaments)
2. Sterile jacket layer   5. Antheridium (*n*)
3. Male gametophyte (*n*)   6. Stalk

Figure 6.22 A longitudinal diagram through
the capsule of a moss. The sporogenous tissue
surrounding columella produces meiospores.

Psilotophyta — whisk ferns

Figure 6.23 The whisk fern, *Psilotum nudum,* is a simple vascular plant lacking true leaves and roots.

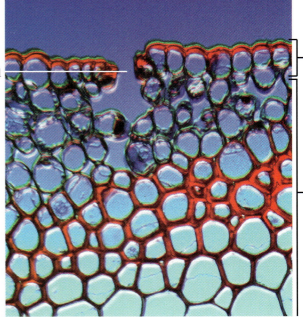

Figure 6.24 A photomicrograph of a scalelike outgrowth from the branch of the whisk fern *Psilotum nudum.*

430X

1. Stoma   2. Epidermis   3. Ground tissue

Figure 6.25 A scanning electron micrograph of a ruptured synangium (3 fused sporangia) of *Psilotum,* which is spilling spores.

75X

1. Sporangium   2. Branch
   (often called asynagium)   3. Spores

Figure 6.26 An aerial stem of the whisk fern *Psilotum nudum.* (a) A cross section and (b) a magnified view of the stele.

1. Stele
2. Cortex
3. Phloem
4. Xylem
5. Epidermis
6. Stoma

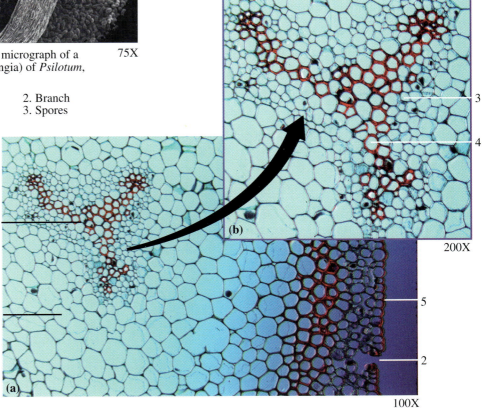

200X

100X

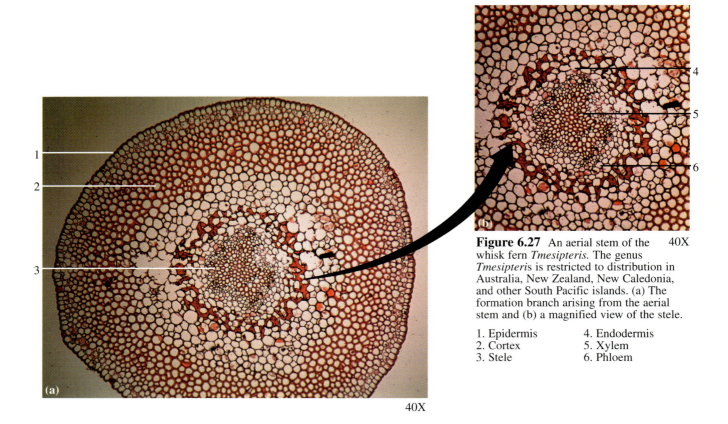

Figure 6.27 An aerial stem of the   40X
whisk fern *Tmesipteris*. The genus
Tmesipteris is restricted to distribution in
Australia, New Zealand, New Caledonia,
and other South Pacific islands. (a) The
formation branch arising from the aerial
stem and (b) a magnified view of the stele.

1. Epidermis   4. Endodermis
2. Cortex   5. Xylem
3. Stele   6. Phloem

40X

Lycophyta — club mosses, quillworts, and spike mosses

Figure 6.28 A club moss *Lycopodium*. Club mosses
occur from the arctic to the tropics. Being evergreen, they

1. Strobilus
2. Leaves (microphylls)
3. Aerial stem

Figure 6.29 An enlargement of an herbarium specimen of
Lycopodium, showing branch tip with sporangia.

1. Sporangia
2. Sporophylls (leaves with attached sporangia)

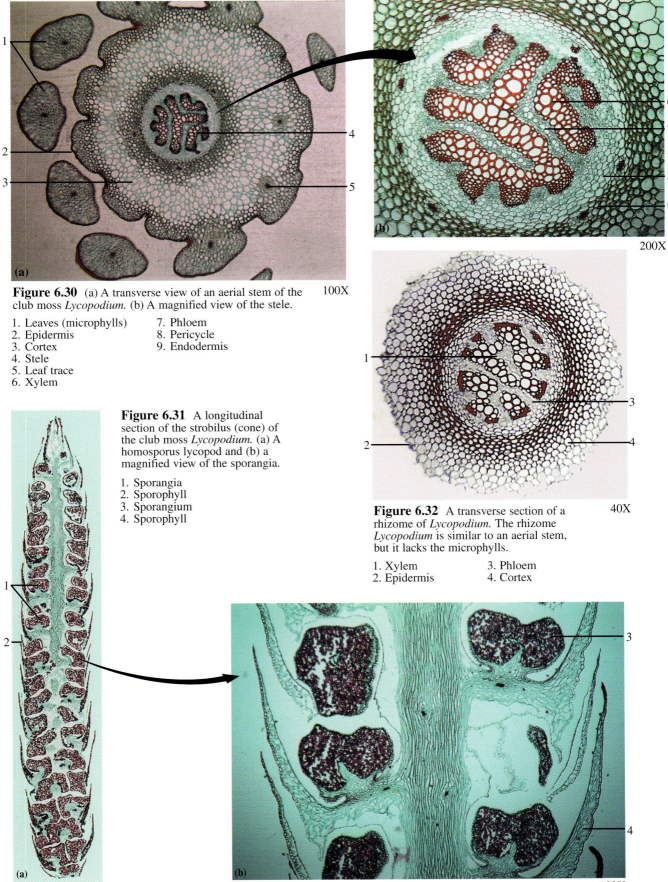

Figure 6.30 (a) A transverse view of an aerial stem of the club moss *Lycopodium.* (b) A magnified view of the stele. 100X

1. Leaves (microphylls)
2. Epidermis
3. Cortex
4. Stele
5. Leaf trace
6. Xylem
7. Phloem
8. Pericycle
9. Endodermis

200X

Figure 6.31 A longitudinal section of the strobilus (cone) of the club moss *Lycopodium.* (a) A homosporus lycopod and (b) a magnified view of the sporangia.

1. Sporangia
2. Sporophyll
3. Sporangium
4. Sporophyll

Figure 6.32 A transverse section of a rhizome of *Lycopodium.* The rhizome *Lycopodium* is similar to an aerial stem, but it lacks the microphylls. 40X

1. Xylem
2. Epidermis
3. Phloem
4. Cortex

15X

100X

Figure 6.34
A young
sporophyte of
Selaginella
growing from a
megagametophyte
inside a megaspore
wall.

1. Root
2. First leaves
3. Stem

40X

Figure 6.33 A herbarium specimen of *Selaginella*. These lycopods are mainly tropical in distribution. Some are found in arid regions, however, where they are dormant during dry seasons.

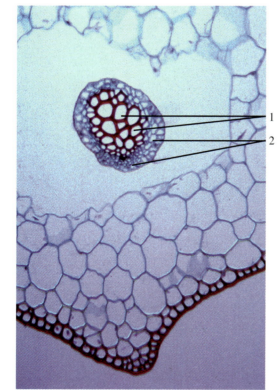

Figure 6.35. A cross section of the   150X
vascular tissue in the stele of *Selaginella*.

1. Xylem   2. Phloem

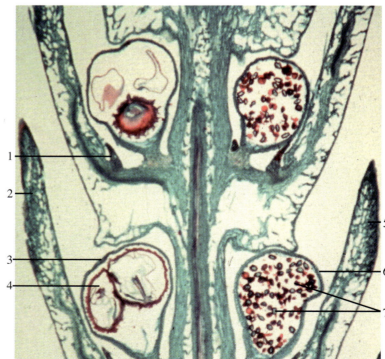

Figure 6.36 A longitudinal section of a strobilus of *Selaginella*.   40X

1. Ligule	4. Megaspore (n)	7. Microspores (n)
2. Megasporophyll	5. Microsporophyll	
3. Megasporangium	6. Microsporangium	

Figure 6.37 A herbarium specimen of a quillwort, *Isoetes melanopoda.* The aquatic quillwort has a small underground stem and quill-like leaves. It is heterosporous and the megasporangia and microsporangia are located at the bases of different leaves.

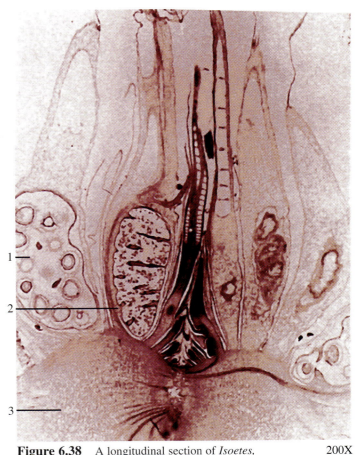

Figure 6.38 A longitudinal section of *Isoetes,*   200X
an aquatic, grass-like lycopod.

1. Megasporangium   3. Corm
2. Microsporangium

Sphenophyta — horsetails

Figure 6.39 The horsetail *Equisetum.* Numerous species of *Sphenophyta* were abundant throughout tropical regions during the Palezoic Era, some 300 million years ago. Currently, *Sphenophyta* are represented by this single genus. Notice the fusing of leaves forming characteristic leaf sheaths. The meadow horsetail, *Equisetum,* showing (a) a section of the stem, (b) a mature strobilus, and (c) a strobilus shedding its spores.

1. Stem
2. Whorl of leaves
3. Separated sporangiophores revealing sporangia
4. Sporangia shedding spores

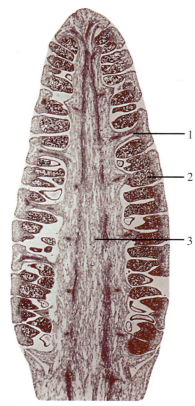

Figure 6.40 A longitudinal   2X
section of the strobilus of
Equisetum.

1. Sporangiophore   3. Strobilus axis
2. Sporangium

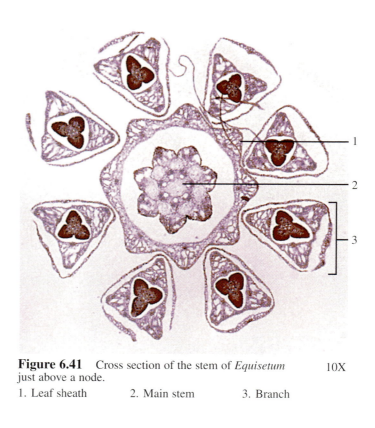

Figure 6.41 Cross section of the stem of *Equisetum*   10X
just above a node.

1. Leaf sheath   2. Main stem   3. Branch

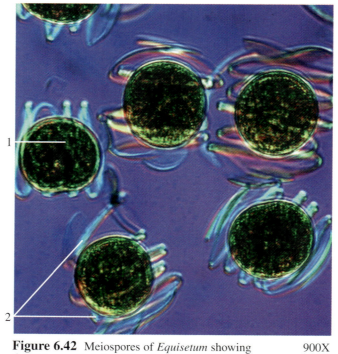

Figure 6.42 Meiospores of *Equisetum* showing   900X
the elaters coiled about them.

1. Meiospore   2. Elater

Pterophytea — ferns

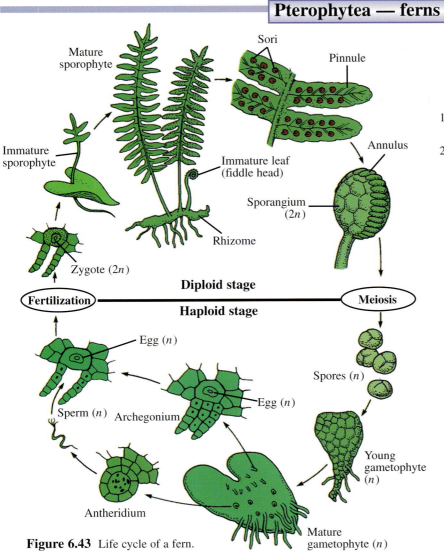

Mature sporophyte

Sori

Pinnule

Immature sporophyte

Immature leaf (fiddle head)

Annulus

Sporangium (2*n*)

Rhizome

Zygote (2*n*)

Diploid stage

Fertilization — Meiosis

Haploid stage

Egg (*n*)

Spores (*n*)

Egg (*n*)

Sperm (*n*)   Archegonium

Young gametophyte (*n*)

Antheridium

Mature gametophyte (*n*)

Figure 6.43   Life cycle of a fern.

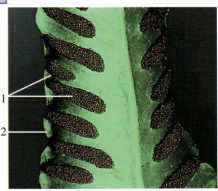

Figure 6.44   The fern, *Asplenium*, showing the sori on the undersurface of the pinna.

1. Sori   2. Pinna

Figure 6.45   The fern, *Cyrtomium*, showing sori on the underside of the pinnae.

1. Pinna   2. Sori

Figure 6.46   The water fern, *Azolla,* is a floating fresh-water plant found throughout Europe and the United States. *Azolla* may become bright orange or red during fall.

Figure 6.47   *Asplenium,* a fern which occurs throughout the Northern Hemisphere.

1. Immature leaf (fiddle head)

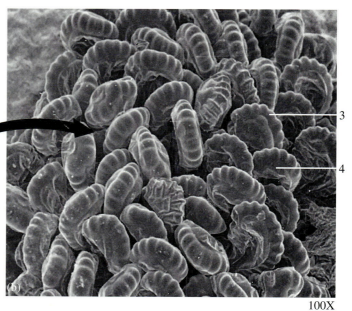

Figure 6.48 The fern *Polypodium.* (a) Sori on the
undersurface of the pinnae, and (b) a scanning electron
micrograph of a sorus.

1. Pinna   3. Annulus
2. Sori   4. Sporangium

100X

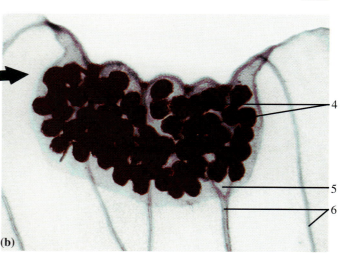

Figure 6.49 The maidenhair fern *Adiantum.* (a) Pinnae and sori and (b) a magnified view of the tip of a pinna folded   100X
under to form a false indusium that encloses the sorus.

1. Sori   3. Pinna   5. False indusium enclosing a sorus
2. False indusium   4. Sporangia with spores   6. Vascular tissue of the pinna

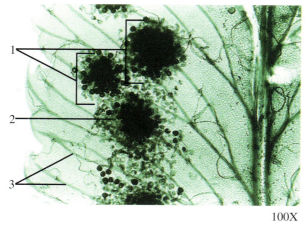

100X

1. Sori   2. Sporangia   3. Veins of pinna

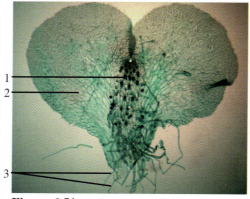

Figure 6.51 A fern gametophyte   40X
showing archegonia.

1. Archegonia   3. Rhizoids
2. Gametophyte
   (prothallus)

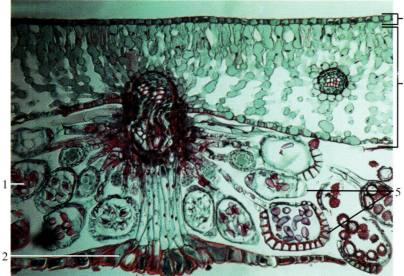

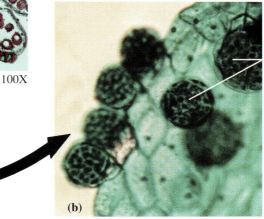

Figure 6.52 A sorus of a homosporous fern, *Cyrtomium falcatum*.

1. Spores (*n*)
2. Indusium (2*n*)
3. Epidermis of pinna
4. Pinna tissue
5. Sporangia (2*n*)

100X

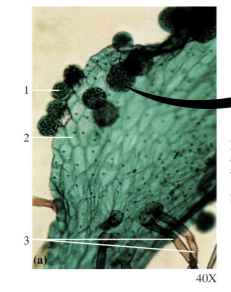

40X

100X

Figure 6.53 (a) A fern gametophyte showing antheridia and (b) a magnified view of sperm within antheridia.

1. Antheridia   3. Rhizoids
2. Gametophyte (prothallus)   4. Sperm within antheridia

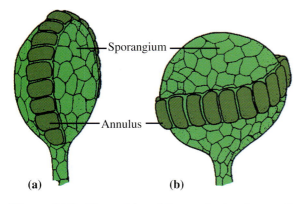

Figure 6.54 The position of the annulus is a feature of importance in fern classification. The sporangium depicted in (a) has a vertical annulus while the sporangium depicted in (b) has a horizontal annulus.

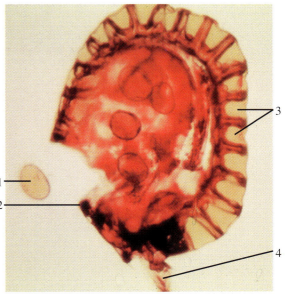

Figure 6.55 The sporangium of the fern   430X
Cyrtomium, discharging a spore.

1. Spore (*n*)   3. Annulus
2. Lip cell   4. Stalk

Cycadophyta— cycads

Figure 6.56   Microsporangiate cones of *Cycas revoluta.*
Cycads are a group of gymnosperms that were very abundant
during the Mesozoic Era. Currently, there are 10 living genera
with about 100 species that are mainly found in tropical and
subtropical areas. The trunk of many cycads is densely covered
with pedioles of shed leaves.

1. Cones

Figure 6.57   An ovulate plant of *Cycas* showing seeds on the upper
surfaces of the megasporaphylls.

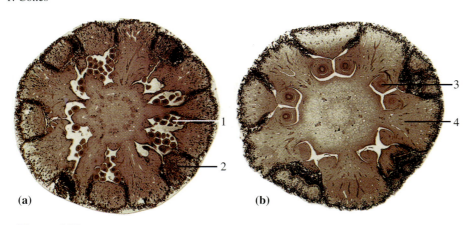

(a)   (b)

Figure 6.58   Cross sections of (a) a microsporangiate cone and (b) a megasporangiate cone
of the cycad *Zamia.*

1. Microsporangia   2. Microsporophyll   3. Ovule   4. Megasporophyll

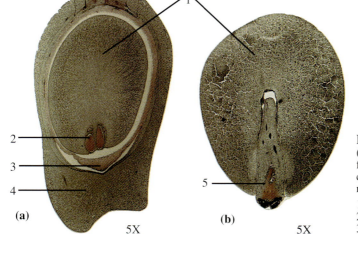

(a)   (b)

5X   5X

Figure 6.59   A microsporangiate
cone of the cycad, *Zamia.*

Figure 6.60   The ovule of the cycad *Zamia.* In
(a) the ovule has two archegonia and is ready to be
fertilized. In (b) the ovule has been fertilized and
contains an embryo, but the seed coat has been
removed.

1. Female gametophyte   4. Integument
2. Archegonium   5. Embryo
3. Megasporangium (nucellus)

Ginkgophyta — *Ginkgo*

Figure 6.61 The ginkgo, or maidenhair tree, *Ginkgo biloba.* Consisting of a central trunk with lateral branches, a mature ginkgo grows to 80 to 100 feet tall. Native to China, *Ginkgo biloba* has been introduced in temperate climates throughout the world as an interesting ornamental tree.

Figure 6.62 A leaf from the ginkgo tree. Notice the characteristic pattern of venation and fan-shaped leaf.

Figure 6.63 A branch of a ginkgo tree, *Ginkgo biloba.*

1. Long shoots   2. Short shoots (spurs)

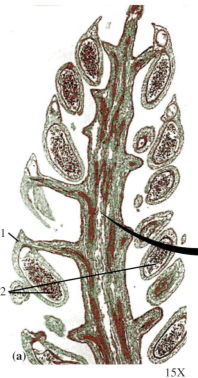

(a)   15X

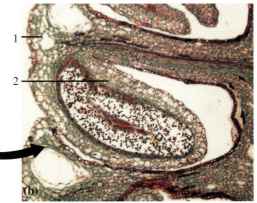

(b)

Figure 6.64 Microsporangiate 15X strobilus of the ginkgo tree. (a) A longitudinal section and (b) a magnified view showing the microsporangia.

1. Sporophyll   2. Microsporangia

Figure 6.65 Male strobili of the ginkgo tree, *Ginkgo biloba.*

1. Leaf   2. Male strobili

Figure 6.66 A mature seed of *Ginkgo.*

Figure 6.67 A longitudinal section of an ovule of *Ginkgo* prior to fertilization.

1. Nucellus
2. Integument
3. Pollen chamber
4. Micropyle

Figure 6.68 A longitudinal section of a seed of *Ginkgo.*

1. Embryo
2. Megagametophyte

A Photographic Atlas for the Biology Laboratory

Coniferophyta — conifers

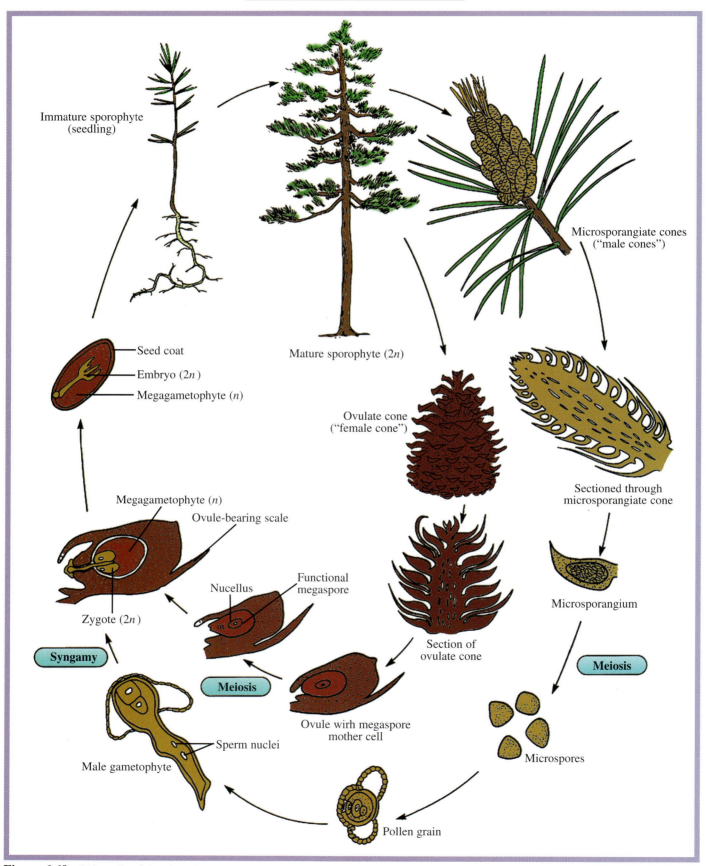

Immature sporophyte (seedling)

Microsporangiate cones ("male cones")

Seed coat

Embryo (2*n*)

Megagametophyte (*n*)

Mature sporophyte (2*n*)

Ovulate cone ("female cone")

Sectioned through microsporangiate cone

Megagametophyte (*n*)

Ovule-bearing scale

Nucellus

Functional megaspore

Zygote (2*n*)

Microsporangium

Syngamy

Meiosis

Section of ovulate cone

Meiosis

Ovule wirh megaspore mother cell

Sperm nuclei

Microspores

Male gametophyte

Pollen grain

Figure 6.69 Life cycle of the pine, *Pinus*.

Figure 6.71 The leaves of most species of conifers are needle-shaped (a) such as these of the blue spruce, *Picea pungens. Araucaria* (b), however, has awl-shaped leaves, and *Podocarpus* (c) has strap-shaped leaves.

Figure 6.70 A young Lodge pole pine, *Contoria picea,* indigenous to the interior western United States.

Courtesy of *Champion Paper Company, Inc.*

Figure 6.72 A diagram of the tissues in the stem (trunk) of a conifer. The periderm (outer bark) protects the tree against water lost and the infestation of insects and fungi. The cells of the phloem (inner bark) compress and become nonfunctional after a relatively short period. The cambium annually produces new bark and xylem and accounts for the growth rings. The secondary xylem is a transporting layer of the stem and provides structural support to the tree.

1. Periderm
2. Phloem
3. Cambium
4. Secondary xylem

Courtesy of Champion Paper Company, Inc.

Figure 6.73 The stem (trunk) of a pine tree that was harvested in 1995 when the tree was 62 years old. The growth rings of a tree indicate environmental conditions that occurred during the tree's life.

1. **1934** — A pine seedling.
2. **1939** — Healthy, undisturbed growth indicated by broad and evenly spaced rings.
3. **1944** — Growth disparity probably due to the falling of a dead tree onto the young healthy six-year old tree. The wider "reaction rings" on the lower side help support the tree.
4. **1954** — The tree is growing straight again, but the narrow rings indicate competition for sunlight and moisture from neighboring trees.
5. **1957** — The surrounding trees are harvested, thus permitting rapid growth once again.
6. **1960** — A burn scar from a fire that quickly scorched the forest.
7. **1972** — Narrow growth rings resulting from a prolonged drought.
8. **1987** — Narrow growth rings, resulting from a sawfly insect infestation, whose larvae eat the needles and buds of many kinds of conifers.

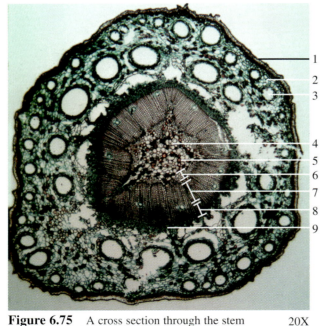

Figure 6.75   A cross section through the stem of a young conifer showing the arrangement of the tissue layers.   20X

1. Periderm
2. Cortex
3. Resin duct
4. Pith
5. Cambium
6. First year's xylem
7. Second year's xylem
8. Third year's xylem
9. Phloem

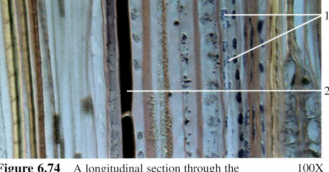

Figure 6.74   A longitudinal section through the phloem of *Pinus*.   100X

1. Sieve areas on a sieve cell
2. Storage parenchyma

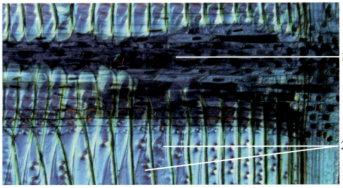

Figure 6.76   A longitudinal section through a stem of *Pinus*, cut through the xylem tissue.   100X

1. Ray parenchyma
2. Tracheids

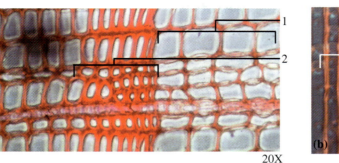

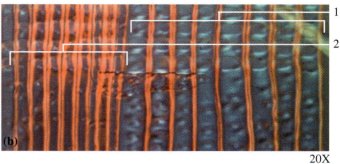

20X   20X

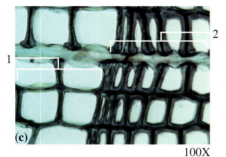

100X

Figure 6.77 Growth rings in *Pinus*. (a) A transverse section through a stem; (b) a radial longitudinal section through a stem; and (c) a magnified view of the xylem cells of spring and summer wood.

1. Spring wood   2. Summer wood

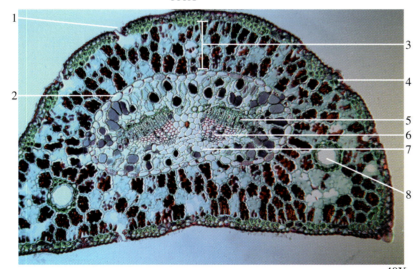

40X

Figure 6.78 A cross section of a leaf (needle) of *Pinus*.

1. Stoma
2. Endodermis
3. Photosynthetic mesophyll
4. Epidermis
5. Phloem
6. Xylem
7. Transfusion tissue
8. Resin duct

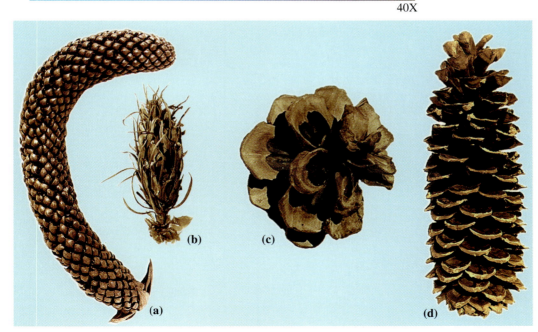

Figure 6.79 Megasporangiate cones from various species of conifers. (a) Southern hemisphere pine, *Aruacaria*; (b) larch, *Larix*; (c) pinion pine, *Pinus edulis*; and (d) sugar pine, *Pinus lambertiana*.

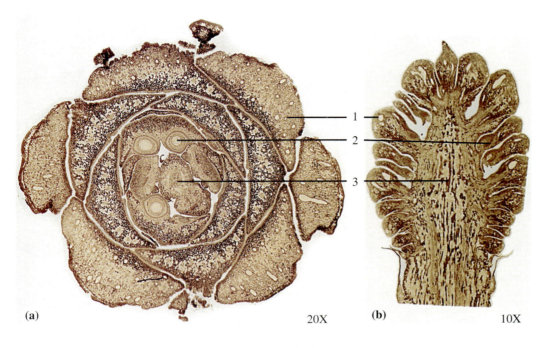

(a) 20X   **(b)** 10X

Figure 6.80 Female cones of a conifer.

1. Ovuliferous scale (megasporophyll)
2. Ovule
3. Cone axis

(a)

Figure 6.81 Microsporangiate cones of a conifer. (a) Staminate cones at end of branch; (b) a longitudinal section through a branch tip; (c) a longitudinal section through a single cone; and (d) a cross section through a single cone.

1. Pine needles (leaves)
2. Sporophylls
3. Staminate cone
4. Microsporangium
5. Sporophyll
6. Cone axis
7. Branch tip

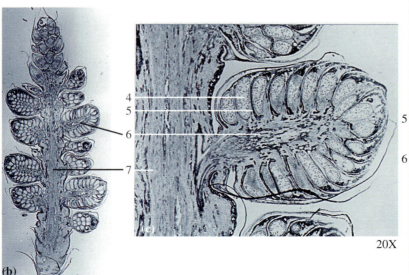

(b) 10X   (c) 20X

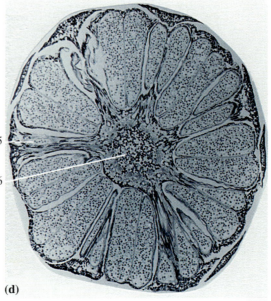

(d) 40X

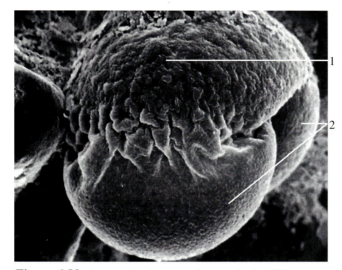

Figure 6.82 A scanning electron micrograph of a *Pinus* pollen grain with inflated bladder-like wings.

1. Pollen body
2. Wings

Figure 6.83 A young sporophyte (seedling) of a pine, *Pinus.*
1. Seedling needles   2. Seed coat   3. Cotyledons

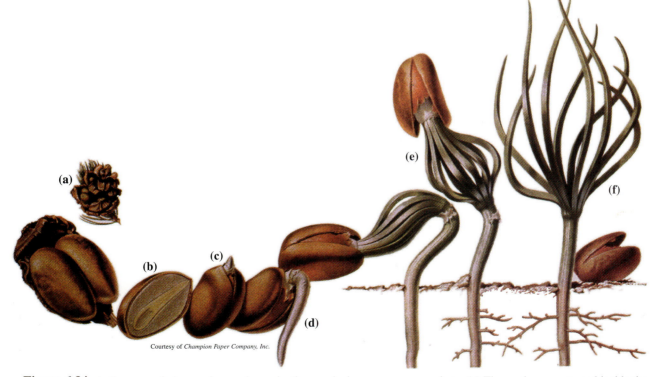

Courtesy of *Champion Paper Company, Inc.*

Figure 6.84 A diagram of pinyon pine seed germination producing a young sporophyte. (a) The seeds are protected inside the cone, two seeds formed on each scale. (b) A sectioned seed showing an embryo embedded in the endosperm. (c) The growing embryo splits the shell of the seed, enabling the root to grow toward the soil. (d) As soon as the tiny root tip penetrates and anchors into the soil, water, and nutrients are absorbed. (e) The leaves (needles) emerge from the shell and create a supply of chlorophyll. Now the sporophyte can manufacture its own food from water and nutrients in the soil and carbon dioxide in the air. (f) Growth occurs at the terminal buds at the base of the leaves.

Anthophyta — angiosperms (moncots and dicots)

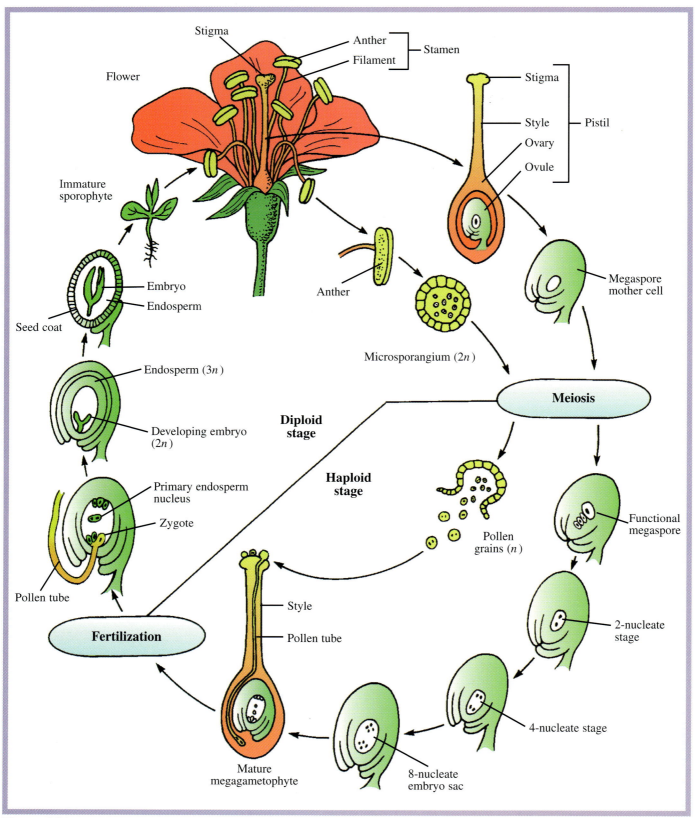

Figure 6.85 Life cycle of an angiosperm.

Figure 6.86 Comparison and examples of monocots and dicots.

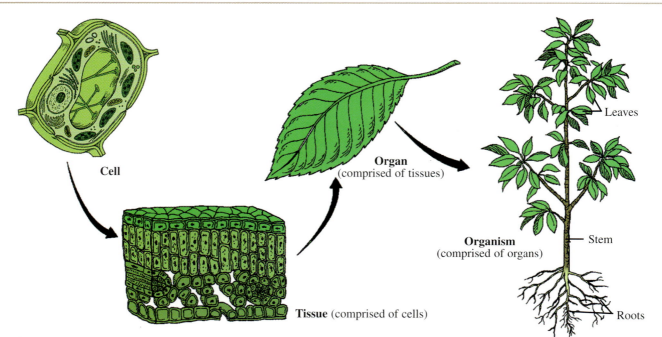

Cell

Organ (comprised of tissues)

Tissue (comprised of cells)

Organism (comprised of organs)

Leaves

Stem

Roots

Figure 6.87 Comparison and examples of monocots and dicots.

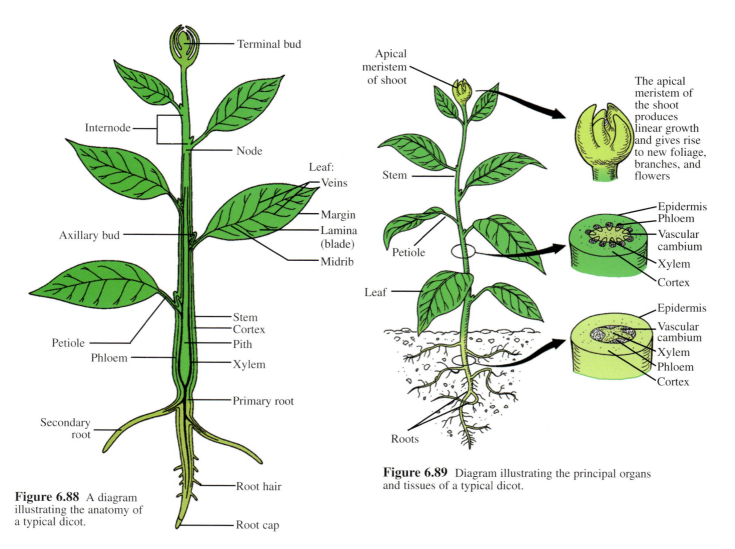

Terminal bud

Internode

Node

Axillary bud

Petiole

Phloem

Secondary root

Stem
Cortex
Pith
Xylem

Primary root

Root hair

Root cap

Leaf:
Veins

Margin
Lamina (blade)
Midrib

Figure 6.88 A diagram illustrating the anatomy of a typical dicot.

Apical meristem of shoot

Stem

Petiole

Leaf

Roots

The apical meristem of the shoot produces linear growth and gives rise to new foliage, branches, and flowers

Epidermis
Phloem
Vascular cambium
Xylem
Cortex

Epidermis
Vascular cambium
Xylem
Phloem
Cortex

Figure 6.89 Diagram illustrating the principal organs and tissues of a typical dicot.

Roots of angiosperms

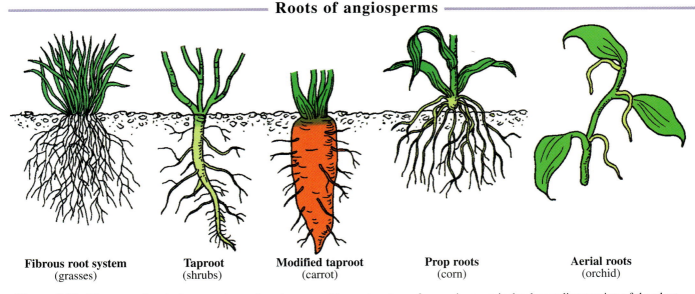

Fibrous root system (grasses)   **Taproot** (shrubs)   **Modified taproot** (carrot)   **Prop roots** (corn)   **Aerial roots** (orchid)

Figure 6.90 Diagrams of typical root systems of angiosperms. The root system of an angiosperm is the descending portion of the plant specialized for anchorage and absorption, storage, and conduction of water and nutrients. Monocots, such as grasses, typically have fibrous root systems. Dicots, such as shrubs and most woody plants, typically have taproot systems. Specialized supporting root systems include prop roots and aerial roots. Taproots, such as found in carrots and turnips, are capable of storing large amounts of food.

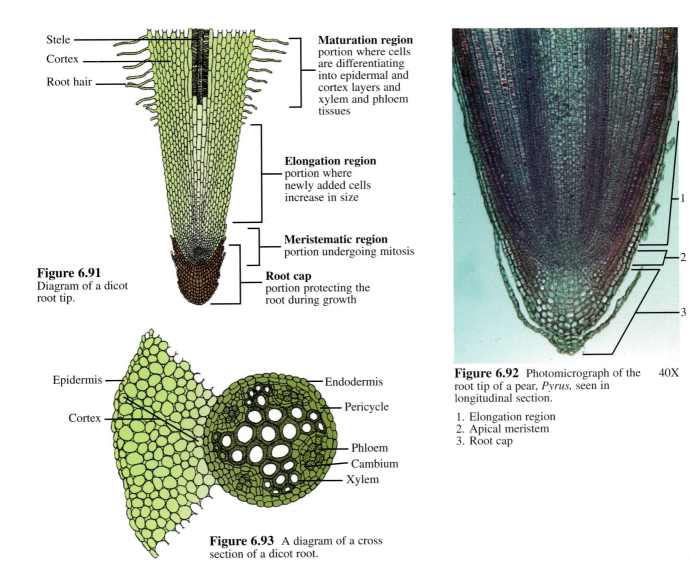

Stele

Cortex

Root hair

Maturation region portion where cells are differentiating into epidermal and cortex layers and xylem and phloem tissues

Elongation region portion where newly added cells increase in size

Meristematic region portion undergoing mitosis

Root cap portion protecting the root during growth

Figure 6.91 Diagram of a dicot root tip.

Epidermis

Cortex

Endodermis

Pericycle

Phloem

Cambium

Xylem

Figure 6.93 A diagram of a cross section of a dicot root.

Figure 6.92 Photomicrograph of the   40X root tip of a pear, *Pyrus,* seen in longitudinal section.

1. Elongation region
2. Apical meristem
3. Root cap

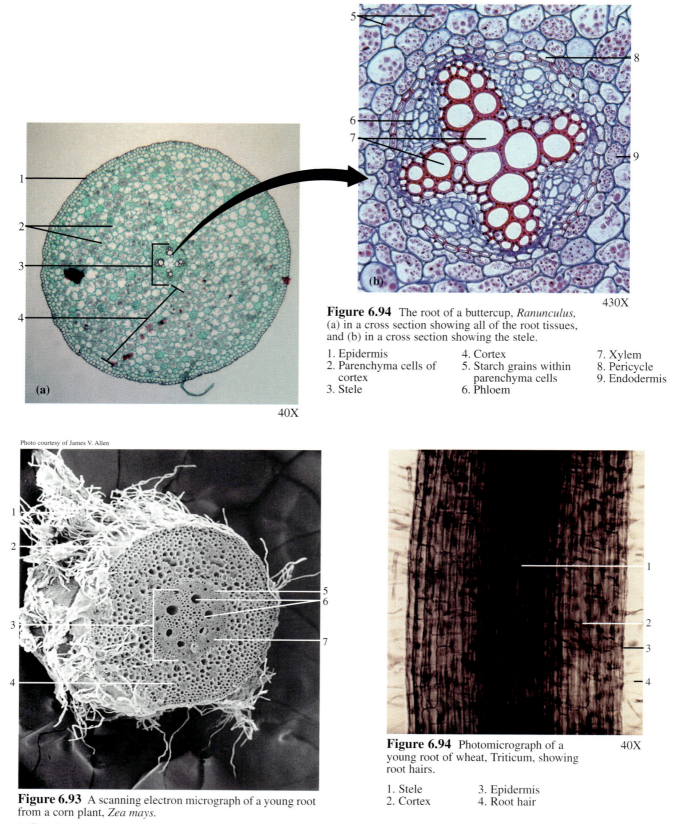

Figure 6.94 The root of a buttercup, *Ranunculus*, (a) in a cross section showing all of the root tissues, and (b) in a cross section showing the stele.

430X

40X

1. Epidermis
2. Parenchyma cells of cortex
3. Stele
4. Cortex
5. Starch grains within parenchyma cells
6. Phloem
7. Xylem
8. Pericycle
9. Endodermis

Photo courtesy of James V. Allen

Figure 6.93 A scanning electron micrograph of a young root from a corn plant, *Zea mays*.

1. Root hairs
2. Epidermis
3. Stele
4. Cortex
5. Endodermis
6. Xylem
7. Phloem

Figure 6.94 Photomicrograph of a young root of wheat, Triticum, showing root hairs.

40X

1. Stele
2. Cortex
3. Epidermis
4. Root hair

Stems of angiosperms

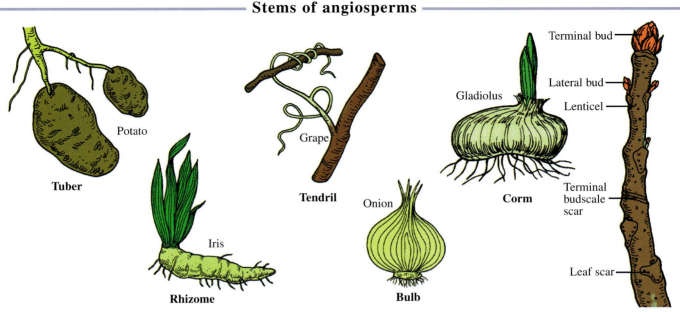

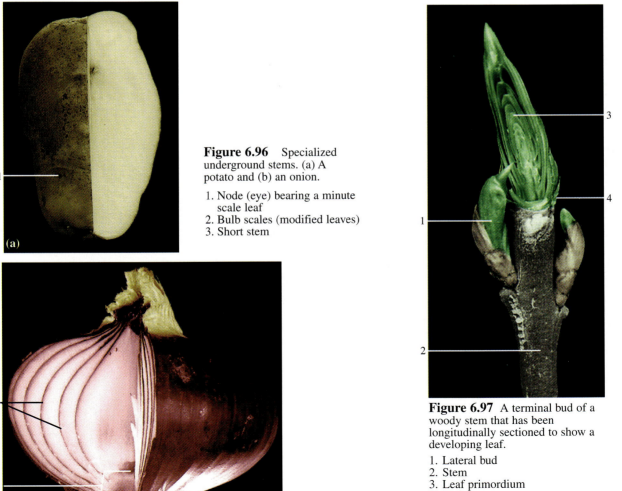

Figure 6.95 Examples of the variety and specialization of angiosperm stems. The stem of an angiosperm is the ascending portion of the plant specialized to produce and support leaves and flowers, transport and store water and nutrients, and provide growth through cell division. Stems of plants are utilized extensively by humans in products including paper, building materials, furniture, and fuel. In addition, the stems of potatoes, onions, celery, cabbage, and other plants are important food crops.

Figure 6.96 Specialized underground stems. (a) A potato and (b) an onion.

1. Node (eye) bearing a minute scale leaf
2. Bulb scales (modified leaves)
3. Short stem

Figure 6.97 A terminal bud of a woody stem that has been longitudinally sectioned to show a developing leaf.

1. Lateral bud
2. Stem
3. Leaf primordium
4. Scale

Figure 6.98 A woody branch of a dicot seen in early spring just as the buds are beginning to swell. Branches and twigs are small extensions of the stems of certain angiosperms, directly supporting leaves and flowers.

1. Terminal bud
2. Internode
3. Terminal budscale scars
4. Lenticel
5. Lateral bud
6. Node area
7. Stem

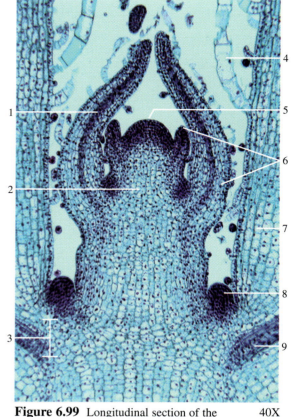

Figure 6.99 Longitudinal section of the   40X
stem tip of the common houseplant *Coleus*.

1. Procambium
2. Ground meristem
3. Leaf gap
4. Trichome
5. Apical meristem
6. Developing leaf primorda
7. Leaf primordium
8. Axillary bud
9. Developing vascular tissue

Figure 6.100 A cross section from the stem of a monocot, *Zea mays*, corn.

1. Phloem
2. Epidermis
3. Vascular bundles
4. Parenchyma cells (ground tissue)
5. Xylem

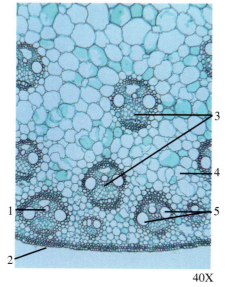

40X

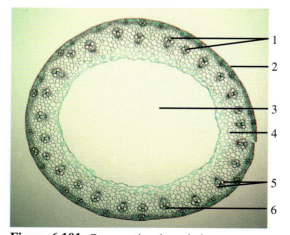

Figure 6.101 Cross section through the stem   40X
of a monocot, *Triticum*, wheat.

1. Vascular bundles of ground tissue
2. Epidermis
3. Ground tissue cavity
4. Parenchyma cells
5. Xylem
6. Phloem

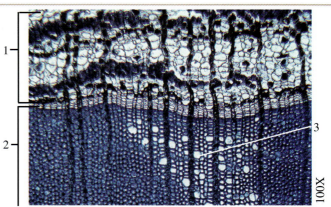

Figure 6.103 A cross section through the secondary xylem (wood) of the stem of an oak, *Quercus.*
1. Summer wood   3. Vessel element 2. Spring wood

Figure 6.102 The stems of saguaro cacti, *Cereus giganteus.*

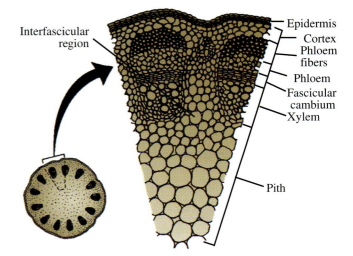

Figure 6.104 A diagram of vascular bundles from the stem of dicot.

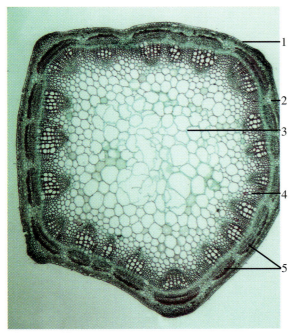

Figure 6.105   A cross section through the stem   40X of a clover.

1. Epidermis   4. Interfascicular region
2. Cortex   5. Vascular bundles with caps
3. Pith

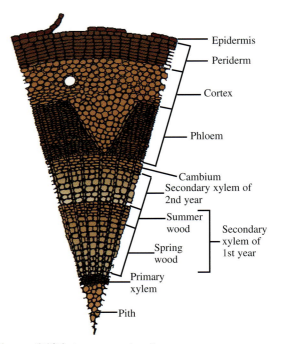

Figure 6.106 A cross section from a two-year stem of linden (basswood).

Courtesy of *Champion Paper Company, Inc.*

Figure 6.107 Samples of bark patterns of representative conifers and angiosperms.

(a) Redwood — The tough, fibrous bark of a redwood tree may be a foot thick. It is highly resistant to fire and insect infestion

(b) Ponderosa pine — The mosaic-like pattern of the bark of mature ponderosa pine is resistant to fire.

(c) White birch — The surface texture of bark on the white birch is like white paper. The bark of the white birch was used by Indians in Eastern United States for making canoes.

(d) Sycamore — The mottled color of the sycamore bark is due to a tendency for large, thin, brittle plates to peel off, revealing lighter areas beneath. These areas grow darker with exposure, until they, too, peel off.

(e) Mangrove — The leathery bark of a mangrove tree is adaptive to brackish water in tropical or semi-tropical regions

(f) Shagbark hickory — The strips of bark in a mature shagbark hickory tree gives this tree its common name.

Leaves of angiosperms

Venation   **Margins**   **Complexity**   **Arrangement on stems**

Figure 6.108 Several representative angiosperm leaf types. Leaves comprise the foliage of plants which provide habitat and a food source for many animals including humans. Leaves also provide protective ground cover and are the portion of the plant most responsible for oxygen replenishment into the atmosphere.

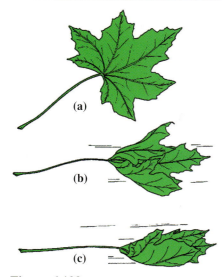

Figure 6.110 A typical angiosperm leaf showing characteristic surface features. Leaves are organs modified to carry out photosynthesis. Photosynthesis is the manufacture of food (sugar) from carbon dioxide and water, with sunlight as the source of energy.

1. Lamina (blade)
2. Petiole
3. Serrate margin
4. Midrib
5. Veins

Figure 6.109 The shape of the leaf (a) is of adaptive value to withstand wind. As the speed of the wind increases (b) and (c), the leaf rolls into a tight cone shape, avoiding damage.

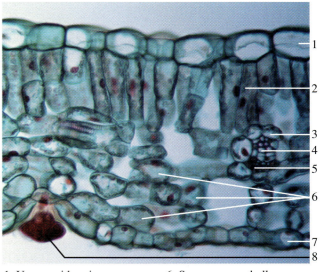

Figure 6.111 The organic decomposition of a leaf is a gradual process beginning with the softer tissues of the lamina, leaving only the vascular tissues of the midrib and the veins, as seen in this photograph. With time, these will also decompose.

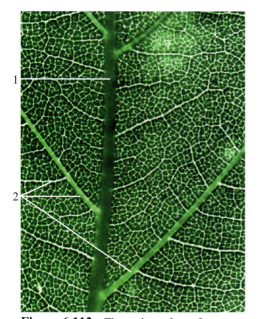

Figure 6.112 The undersurface of an angiosperm leaf showing the vascular tissue lacing through the lamina, or blade, or the leaf.

1. Midrib   2. Veins

Figure 6.114 A cross section through the leaf of the common hedge privet *Ligustrum*. The typical tissue arrangement of a leaf includes an upper epidermis, a lower epidermis, and the centrally located mesophyll. Containing chloroplasts, the cells of the mesophyll are often divided into palisade mesophyll and spongy mesophyll. Veins within the mesophyll conduct material through the leaf.

1. Upper epidermis
2. Palisade mesophyll
3. Bundle sheath
4. Xylem
5. Phloem

6. Spongy mesophyll
7. Lower epidermis
8. Trichome (leaf hair)

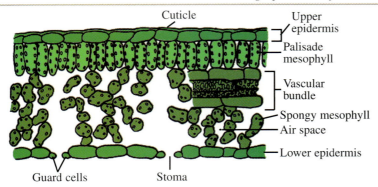

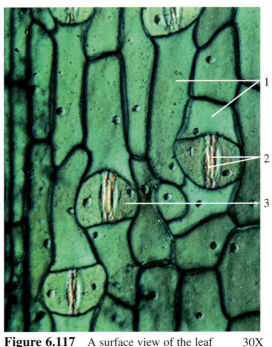

Figure 6.115   A cross sectional diagram of a dicot leaf.

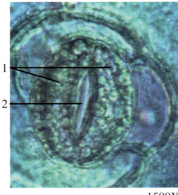

1500X

Figure 6.116   Guard cells of and stomata of a geranium. Guard cells regulate the opening of the stomata according to the environmental humidity.

1. Guard cells
2. Stomata

Figure 6.117   A surface view of the leaf   30X
epidermis of *Trandescantia*.

1. Epidermal cells   3. Subsidiary cells
2. Guard cells surrounding stomata

Flower of angiosperms

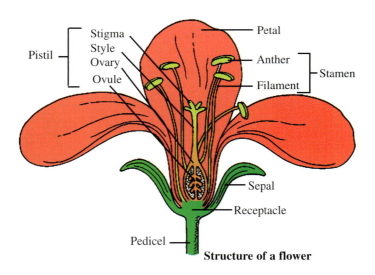

Structure of a flower

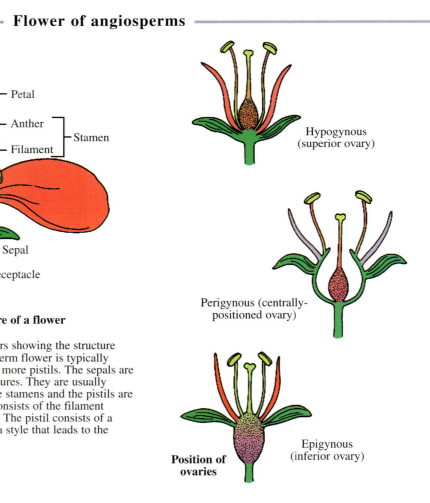

Hypogynous (superior ovary)

Perigynous (centrally-positioned ovary)

Epigynous (inferior ovary)

Position of ovaries

Figure 6.118   Diagrams of angiosperm flowers showing the structure and relative position of the ovaries. The angiosperm flower is typically composed of sepals, petals, stamens, and one or more pistils. The sepals are the outermost circle of protective leaf-like structures. They are usually green and are collectively called the corolla. The stamens and the pistils are the reproductive parts of the flower. A stamen consists of the filament (stalk) and the anther, where pollen is produced. The pistil consists of a sticky stigma at the tip that receives pollen and a style that leads to the ovary.

Amaryllis – *Amaryllis bellandonna*

Onagraceae – *Oenothera caepitosa*

Onagraceae – *Fuschsia*

Monotropaceae
Oenothera caepitosa

Ranunculaceae – *Aquilegia coerulea*

Cactaceae – *Opuntia polyacantha*

Figure 6.119 A variety of angiosperm flowers.

Figure 6.120 The floral structure of a tulip.

1. Stigma   3. Petal
2. Anther   4. Ovary

Figure 6.121 The structure of a dissected *Pyrus*, showing an epigyynous flower.

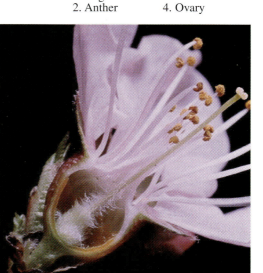

Figure 6.122 The structure of a dissected rose, *Prunus,* showing a perigynous flower.

(a)

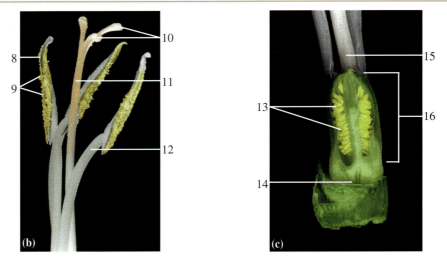

(b)   (c)

Figure 6.123   (a) The floral structure of *Gladiolus.* (b) The anthers and stigma and (c) the ovary.

1. Anther	7. Ovary	13. Ovules
2. Filament	8. Anther	(unfertilized seeds)
3. Ovules	9. Pollen (*n*)	14. Receptacle
4. Receptacle	10. Stigma	15. Style
5. Stigma	11. Style	16. Ovary
6. Style	12. Filament	

Figure 6.123   A dissected cactus flower, *Echinocereus,* showing an inferior ovary with numerous stamens.

Figure 6.124   A dissected rose, *Chaenomeles japonica,* showing an epigynous flower.

Figure 6.125   The floral parts of a grass, *Elymus flavescens,* showing spikelets with six florets.

Figure 6.126   The floral structure of grasses.

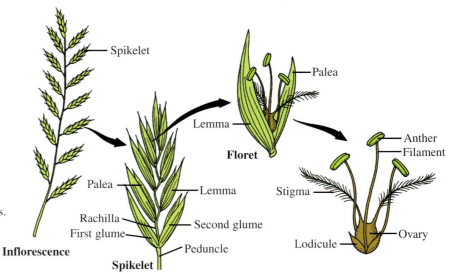

Spikelet

Palea

Lemma

Floret

Palea

Lemma

Rachilla

First glume

Second glume

Peduncle

Inflorescence

Spikelet

Anther

Filament

Stigma

Lodicule

Ovary

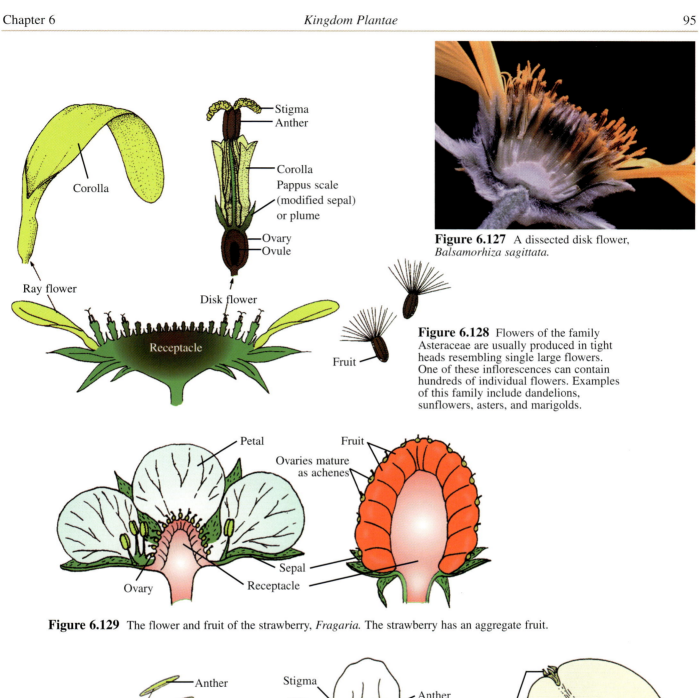

Figure 6.127 A dissected disk flower, *Balsamorhiza sagittata.*

Figure 6.128 Flowers of the family Asteraceae are usually produced in tight heads resembling single large flowers. One of these inflorescences can contain hundreds of individual flowers. Examples of this family include dandelions, sunflowers, asters, and marigolds.

Figure 6.129 The flower and fruit of the strawberry, *Fragaria.* The strawberry has an aggregate fruit.

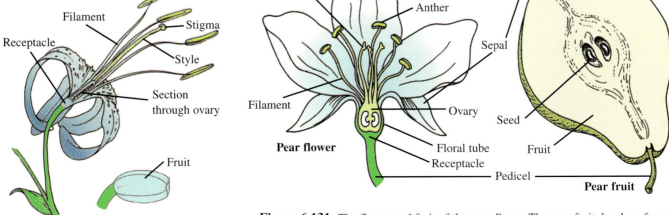

Figure 6.130 The flower and fruit of the lily, *Lilium.*

Figure 6.131 The flower and fruit of the pear *Pyrus.* The pear fruit develops from the floral tube (fused perianth) as well as the ovary.

Seeds, fruits, and seed germination of angiosperms

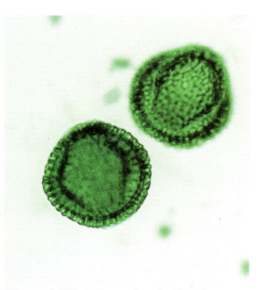

Figure 6.132 Pollen grains of a lilac, *Syringa*.

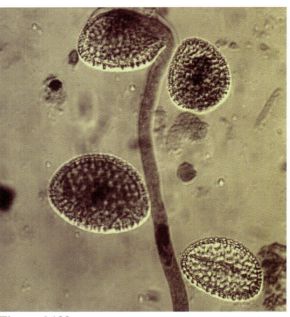

Figure 6.133 Pollen grains of a lily. The pollen grain at the top of the photo has germinated to produce a pollen tube.

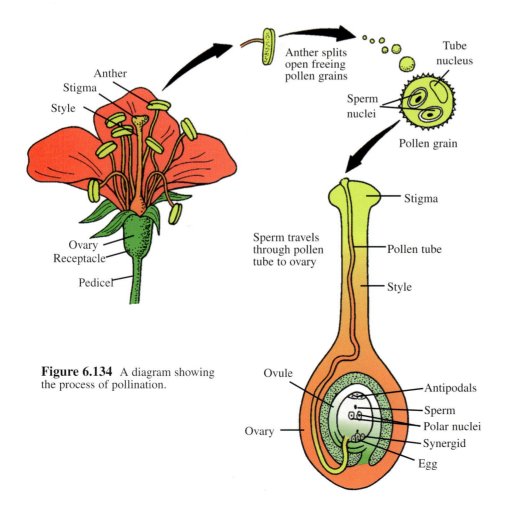

Anther splits open freeing pollen grains

Tube nucleus

Sperm nuclei

Pollen grain

Anther

Stigma

Style

Sperm travels through pollen tube to ovary

Ovary

Receptacle

Pedicel

Stigma

Pollen tube

Style

Ovule

Ovary

Antipodals

Sperm

Polar nuclei

Synergid

Egg

Figure 6.134 A diagram showing the process of pollination.

(a)

(b)

Figure 6.135 The flower (a) and the fruits (b) of the dandelion, *Taraxacum*. The dandelion has a composite flower. The wind-borne fruit (containing one seed) of a dandelion, and many other members of the family Asteraceae, develop a plumlike pappus, which enables the light fruit to float in the air.

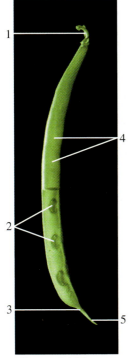

Figure 6.136 A photograph of a dissected legume, string bean.

1. Pedicel
2. Seeds
3. Style
4. Fruit
5. Stigma

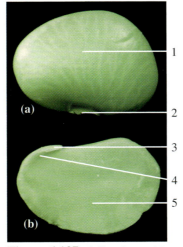

Figure 6.137 A lima bean. (a) The entire bean seed and (b) a longitudinally sectioned seed

1. Integument (seed coat)
2. Hilum
3. Radicle
4. Hypocotyl
5. Cotyledon

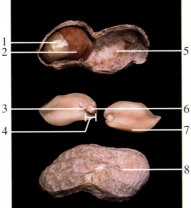

Figure 6.139 The fruit and seeds of a peanut plant.

1. Cotyledon
2. Integument (seed coat)
3. Plumule
4. Embryo
5. Interior of fruit
6. Radicle
7. Cotyledon
8. Fruit wall

Figure 6.138 A photomicrograph of the seed coat of the garden bean, *Phaseolus*, showing the sclerified epidermis   100X

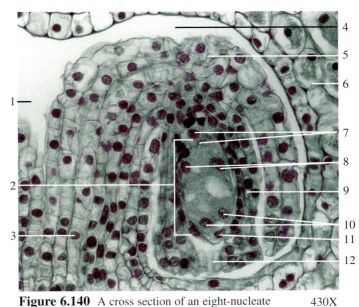

Figure 6.140 A cross section of an eight-nucleate embryo sac of an ovule from a lily, *Lilium*.   430X

1. Locule
2. Embryo sac
3. Funiculus
4. Chalaza
5. Ovule
6. Ovary
7. Antipodal cells
8. Polar nuclei (one is *n*, one is 3*n*)
9. Outer integument (2*n*)
10. Inner integument (2*n*)
11. Synergid cells (*n*)
12. Egg (*n*)
13. Microphyte (pollen tube entrance)

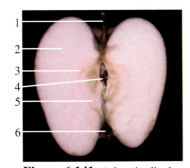

Figure 6.141 A longitudinal section of an apple fruit.

1. Pedicel
2. Receptacle
3. Ovary wall
4. Seed (mature ovule)
5. Mature ovary
6. Remnants of floral parts

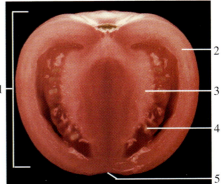

Figure 6.142 A cross section through a grapefruit fruit.

1. Seed
2. Exocarp
3. Mesocarp
4. Endocarp
5. Pericarp

Figure 6.143 A longitudinal section of a tomato fruit (a berry).

1. Mature ovary
2. Ovary wall
3. Placenta
4. Seed
5. Remnant of stigma

Figure 6.144 A longitudinal section of a pineapple fruit.

1. Floral parts
2. Central axis.

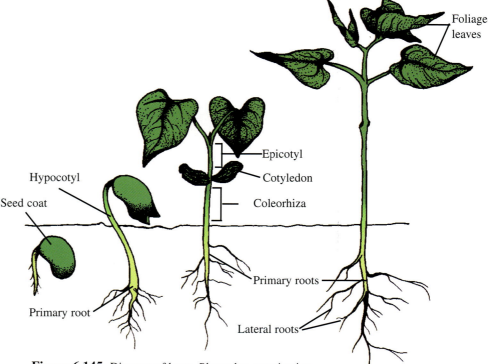

Figure 6.145 Diagram of bean, *Phaseolus,* germination.

Seed coat

Hypocotyl

Primary root

Epicotyl

Cotyledon

Coleorhiza

Foliage leaves

Primary roots

Lateral roots

Courtesy of *Champion Paper Company, Inc.*

(a) (b) (c) (d)

(e) (f) (g) (h)

(i) (j) (k) (l)

Figure 6.146 Some examples of seed dispersal.

(a) Maple — The winged seeds of a maple fall with a spinning motion that may carry it hundreds of yards from the parent tree

(b) White pine — The second-year cones of a white pine open to expose the winged seeds to the wind

(c) Willow — The air-borne seeds of a willow may be dispersed over long distances.

(d) Witch hazel — Mature seeds of the witch hazel tree are dispersed up to 10 feet by forceful discharge

(e) Mangrove — The fruits of this tropical tree begin to germinate while still on the branch, forming pointed roots. When the seeds drop from the tree, they may float to a muddy area where the roots take hold

(f) Coconut — The buoyant, fibrous husk of a coconut permits dispersal from one island or land mass to another by ocean currents.

(g) Pecan — The seed husk of a pecan provides buoyancy and protection as it is dispersed by water.

(h) Black walnut — The encapsulated seed of the black walnut is dispersed through burial by a squirrel or floating in a stream

(i) Apple — The seeds of an apple tree may be dispersed by animals that ingest the fruit and pass the undigested seeds hours later in their feces.

(j) Cherry — Moderate sized birds, such as robins, may carry a ripe cherry to a eating site where the juicy pulp is eaten and the hard seed is discarded

(k) Beech — Seeds from a beech tree are dispersed by mammals as the spiny husks adhere to their hair. In addition, many mammals ingest these seeds and disperse them in their feces.

(l) Oak — An oak seed may be dispersed through burial by a squirrel

Kingdom Animalia

7

Animals are multicellular, heterotrophic eukaryotes that ingest food materials and store carbohydrate reserves as glycogen or fat. The cells of animals lack cell walls, but do contain intercellular connections including desmosomes, gap junctions, and tight junctions. Animal cells are also highly specialized into the specific kinds of tissues described in chapter 1. Most animals are motile through the contraction of muscle fibers containing actin and myosin proteins. The complex body systems of animals include elaborate sensory and neuromotor specializations that accommodate dynamic behavioral mechanisms.

Reproduction in animals is primarily sexual, with the diploid stage generally dominating the life cycle. Primary sex organs, or *gonads*, produce the haploid gametes called *sperm* and *ova*. Propagation begins as a small flagellated sperm fertilizes a larger, nonmotile egg forming a diploid zygote that has genetic traits of both parents. The zygote then undergoes a succession of mitotic divisions called *cleavage*. In most animals, cleavage is followed by the formation of a multicellular stage called a *blastula*. With further development, the *germ layers* form which eventually give rise to each of the body organs. The developmental cycle of many animals includes *larval forms*, which are still developing, free-living, and sexually immature. Larvae usually have different food and habitat requirements from those of the adults. Larvae eventually undergo metamorphosis that transforms them into sexually mature adults.

Animals inhabit nearly all aquatic and terrestrial habitats of the biosphere. The greatest number of animals are marine, where the first animals probably evolved. Depending on the classification scheme, animals may be grouped into as many as 35 phyla. The most commonly known phylum is *Chordata* (table 7.1) that includes the subphylum *Vertebrata*, or the backboned animals. Chordates, however, comprise only about 5% of all the animal species. All other animals are frequently referred to as *invertebrates*, and they account for approximately 95% of the animal species.

Table 7.1 Some Representatives of the Kingdom Animalia

Phyla and Representative Kinds	Characteristics
Porifera — sponges	Multicellular, aquatic animals, with stiff or fibrous bodies perforated by pores
Cnidaria — corals, hydra, jellyfish	Aquatic animals, radially symmetrical, mouth surrounded by tentacles bearing cnidocytes (stinging cells); body composed of epidermis and gastrodermis, separated by mesoglea
Platyhelminthes — flatworms	Elongated, flattened, and bilaterally symmetrical; distinct head containing ganglia; nerve cords; protonephridia of flame cells
Nematoda — roundworms	Mostly microscopic, unsegmented worms; body enclosed in cuticle; whiplike body movement
Mollusca — mollusks: clams, snails, squids	Bilaterally symmetrical with true coelom; mantle; may have muscular foot and protective shell
Annelida — segmented worms	Body segmented by septa; a series of hearts; hydrostatic skeleton and circular and longitudinal muscles
Arthropoda — crustaceans, insects	Body segmented; paired and jointed appendages; chitinous exoskeleton; hemocoel for blood flow
Echinodermata — echinoderms: sea stars, sand dollars, sea cucumbers, sea urchins	Larvae have bilateral symmetry; adults have pentaradial symmetry; coelom with complete digestive tract; regeneration of body parts
Chordata — amphioxus, amphibians, fishes, reptiles, birds, and mammals	Fibrous notochord, pharyngeal gill pouches, and dorsal hollow nerve cord present at some stage in their life cycle

Porifera — sponges

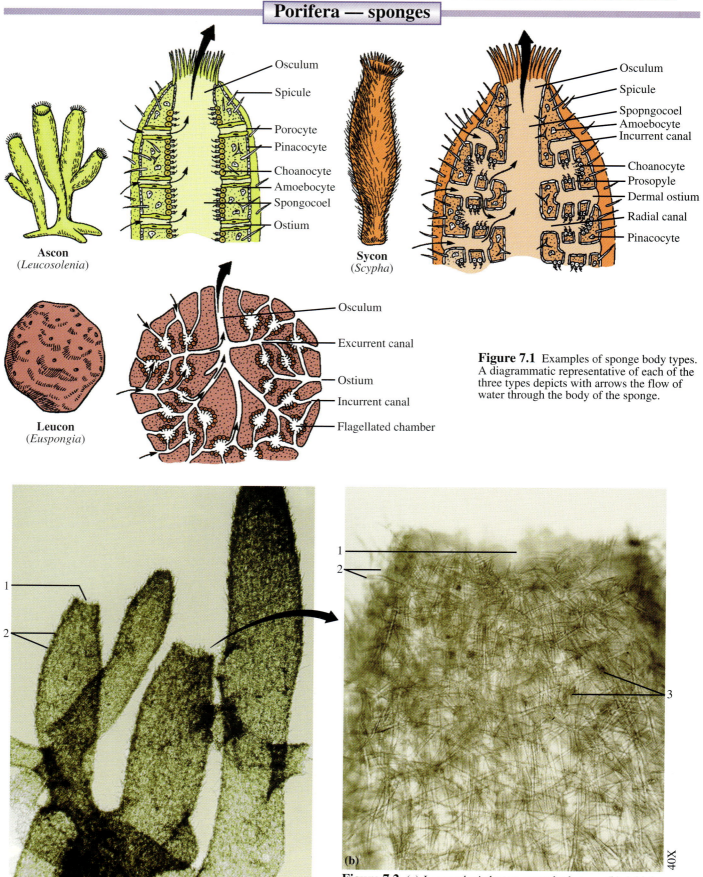

Ascon
(*Leucosolenia*)

Osculum
Spicule
Porocyte
Pinacocyte
Choanocyte
Amoebocyte
Spongocoel
Ostium

Sycon
(*Scypha*)

Osculum
Spicule
Spopngocoel
Amoebocyte
Incurrent canal
Choanocyte
Prosopyle
Dermal ostium
Radial canal
Pinacocyte

Leucon
(*Euspongia*)

Osculum
Excurrent canal
Ostium
Incurrent canal
Flagellated chamber

Figure 7.1 Examples of sponge body types. A diagrammatic representative of each of the three types depicts with arrows the flow of water through the body of the sponge.

Figure 7.2 (a) *Leucosolenia* has an ascon body type. (b) A high magnification of the spicules and ostia.

1. Osculum   2. Spicules   3. Ostia

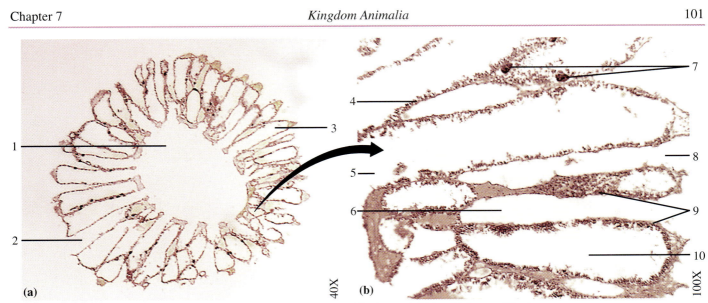

(a) 40X   (b) 100X

Figure 7.3 Cross sections of the sponge, *Scypha.* (a) is a low magnification and (b) is a high magnification.

1. Spongocoel
2. Incurrent canal
3. Radial canal
4. Choanocytes (collar cells)

5. Apopyle
6. Incurrent canal
7. Blastulas
8. Ostium

9. Pinacocytes
10. Radial canal

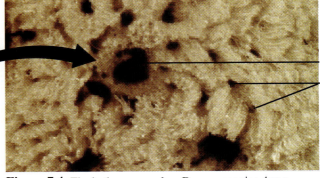

Figure 7.4 The bath sponge, class Desmospongiae, has a leuconoid body structure.

1. Osculum
2. Ostia

Figure 7.5 Spicules of a freshwater sponge.   430X

Figure 7.6 A leucon type sponge occurring in a natural habitat.

Cnidaria — corals, hydra, and jellyfish

Table 7.2 Some Representatives of the Phylum Cnidaria

Classes and Representative Kinds	Characteristics
Hydrozoa — hydra, *Obelia*, Portuguese man-of-war	Mainly marine; both polyp and medusa stage (polyp form only in hydra); polyp colonies in some
Scyphozoa — jellyfish	Marine coastal waters; polyp stage restricted to small larval forms
Anthozoa — sea anemones, colonial corals, sea fans	Marine coastal waters; solitary or colonial polyps; no medusa stage; partitioned gastrovascular cavity

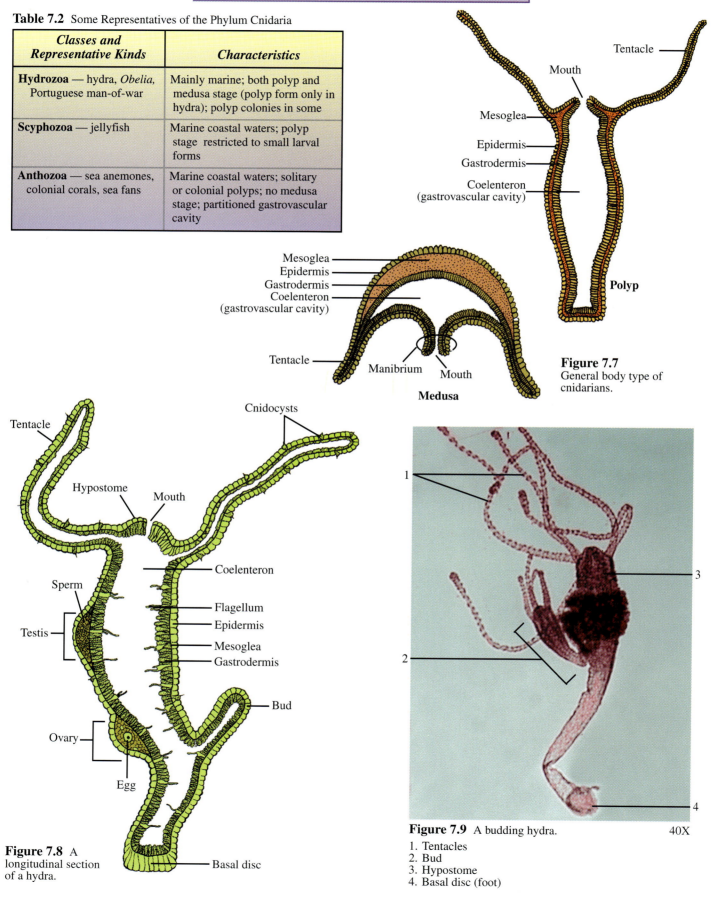

Tentacle

Mouth

Mesoglea

Epidermis

Gastrodermis

Coelenteron (gastrovascular cavity)

Polyp

Mesoglea
Epidermis
Gastrodermis
Coelenteron (gastrovascular cavity)

Tentacle

Manibrium   Mouth

Medusa

Figure 7.7
General body type of cnidarians.

Cnidocysts

Tentacle

Hypostome   Mouth

Coelenteron

Sperm

Testis

Flagellum
Epidermis
Mesoglea
Gastrodermis

Bud

Ovary

Egg

Basal disc

Figure 7.8 A longitudinal section of a hydra.

Figure 7.9 A budding hydra.   40X

1. Tentacles
2. Bud
3. Hypostome
4. Basal disc (foot)

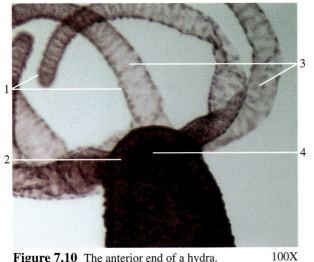

Figure 7.10 The anterior end of a hydra.   100X
1. Cnidoblasts   3. Tentacles
2. Hypostome   4. Mouth

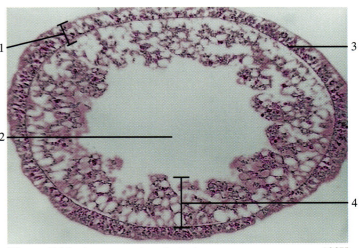

Figure 7.11   A cross section of a hydra.   100X
1. Epidermis (ectoderm)   3. Mesoglea
2. Coelenteron   4. Gastrodermis (endoderm)

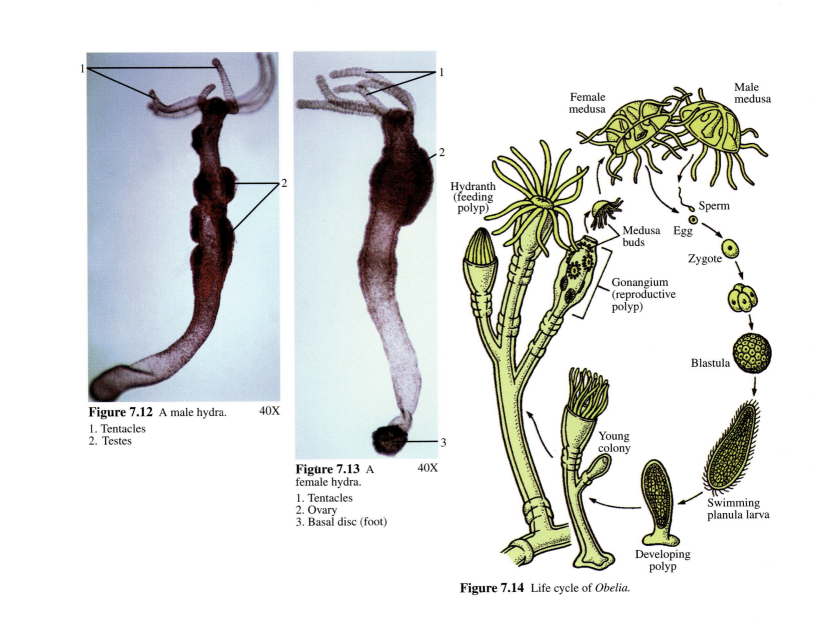

Figure 7.12 A male hydra.   40X
1. Tentacles
2. Testes

Figure 7.13 A   40X
female hydra.
1. Tentacles
2. Ovary
3. Basal disc (foot)

Figure 7.14 Life cycle of *Obelia*.

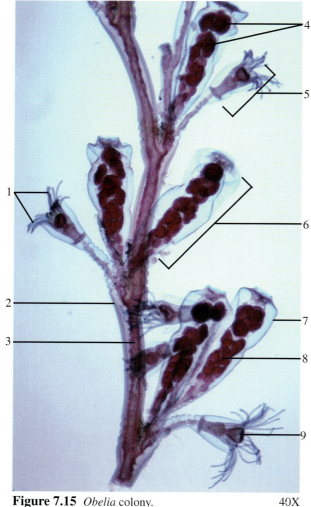

Figure 7.15 *Obelia* colony.   40X

1. Tentacles
2. Perisarc
3. Coenosarc (reproductive polp)
4. Medusa buds
5. Hydranth (feeding polp)
6. Gonangium
7. Gonotheca
8. Blastostyle
9. Hypostome

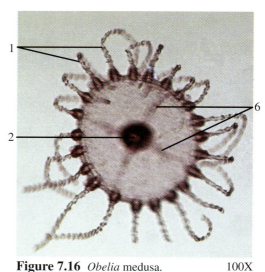

Figure 7.16 *Obelia* medusa.   100X

1. Tentacles   3. Gonads
2. Manubrium

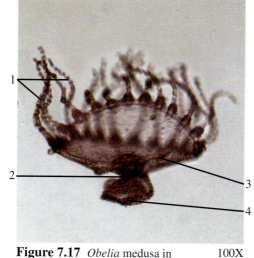

Figure 7.17 *Obelia* medusa in   100X
feeding position.

1. Tentacles   3. Gonad
2. Manubrium   4. Mouth

Figure 7.18 The portugese man-of-war, *Physalia,* is actually a colony of medusae and polyps acting as a single organism. The tentacles are comprised of three types of polyps: the gastrozooids (feeding polyps), the dactylozooids (stinging polyps), and the gonozooids (reproductive polyps).

1. Pneumatophore (float)   2. Tentacles

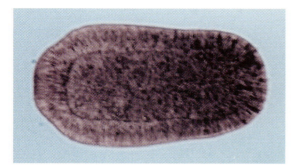

Figure 7.19 The *Aurelia* planula larva develops 40X from a fertilized egg that may be retained on the oral arm of the medusa.

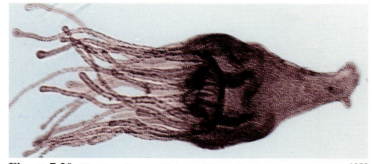

Figure 7.20 The *Aurelia* scyphistoma. The polyp is a 40X developmental stage in the life cycle of the jelly fish.

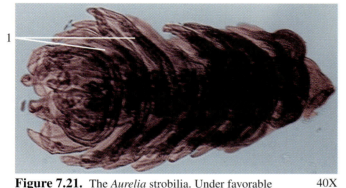

Figure 7.21. The *Aurelia* strobilia. Under favorable 40X conditions, the scyphistoma develops into the strobilia.
1. Developing ephyrae

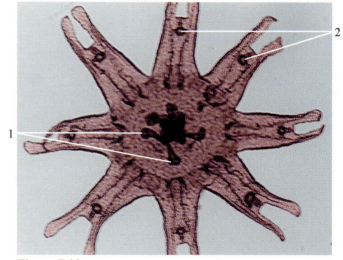

Figure 7.22 The *Aurelia* ephyra larva. These 40X gradually develop into adult jelly fish.
1. Gonads   2. Rhopalia (sense organs)

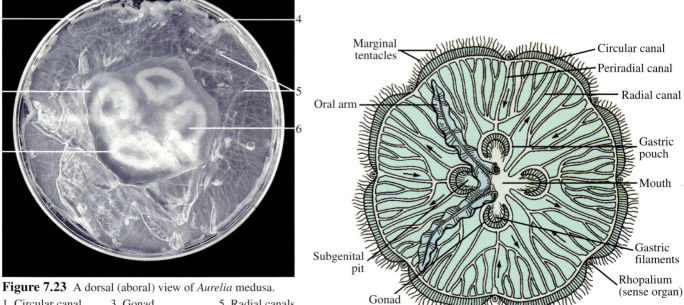

Figure 7.23 A dorsal (aboral) view of *Aurelia* medusa.
1. Circular canal   3. Gonad   5. Radial canals
2. Gastric pouch   4. Marginal tentacles 6. Subgenital pit

Marginal tentacles
Oral arm
Subgenital pit
Gonad

Circular canal
Periradial canal
Radial canal
Gastric pouch
Mouth
Gastric filaments
Rhopalium (sense organ)

Figure 7.24 A ventral (oral) view of *Aurelia* medusa. In this diagram, the right oral arms have been removed. The arrows depict circulation through the canal system.

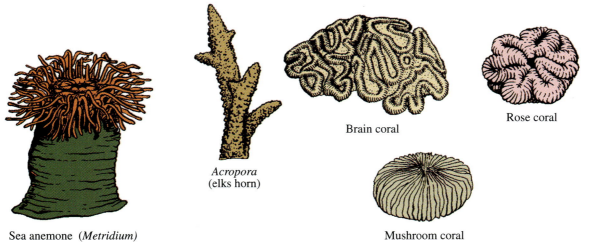

Sea anemone (*Metridium*)

Acropora
(elks horn)

Brain coral

Rose coral

Mushroom coral

Figure 7.25   Representatives of the class Anthozoa.

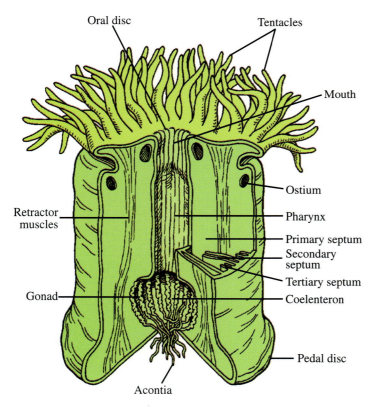

Oral disc

Tentacles

Mouth

Ostium

Pharynx

Primary septum

Secondary septum

Tertiary septum

Coelenteron

Pedal disc

Retractor muscles

Gonad

Acontia

Figure 7.26   A diagram of a partially dissected sea anemone, *Metridium.*

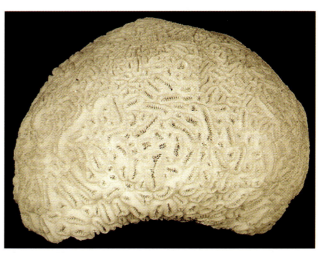

Figure 7.27   Dried skeleton of a brain coral, class Anthozoa.

Figure 7.28   A specimen of coral, class Anthozoa. This type of coral comprises coral reefs.

Platyhelminthes — flatworms

Table 7.3 Some Representatives of the Phylum Platyhelminthes

Classes and Representative Kinds	Characteristics
Turbellaria — planarians	Mostly free-living, carnivorous, freshwater forms; body covered by ciliated epidermis
Trematoda — flukes (schistosomes)	Parasitic with wide range of invertebrate and vertebrate hosts; suckers for attachment to host
Cestoda — tapeworms	Parasitic on many vertebrate hosts; complex life cycle with intermediate hosts; suckers or hooks on scolex for attachment to host; eggs are produced and shed within proglottids

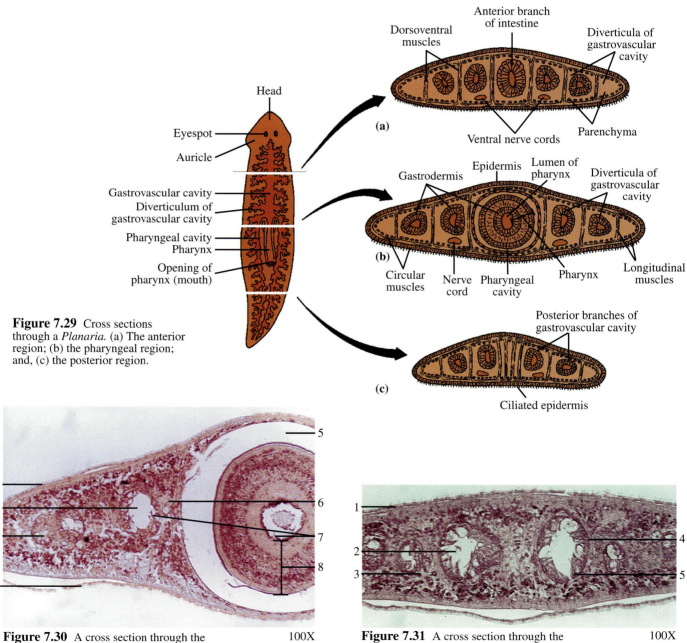

Figure 7.29 Cross sections through a *Planaria*. (a) The anterior region; (b) the pharyngeal region; and, (c) the posterior region.

Figure 7.30 A cross section through the pharyngeal region of *Planaria*.   100X

1. Epidermis
2. Gastrovascular cavity
3. Testis
4. Cilia
5. Pharyngeal cavity
6. Dorsoventral muscles
7. Gastrodermis (endoderm)
8. Pharynx

Figure 7.31 A cross section through the posterior region of *Planaria*.   100X

1. Epidermis
2. Gastrovascular cavity
3. Mesenchyme
4. Dorsoventral muscles
5. Intestine (endoderm)

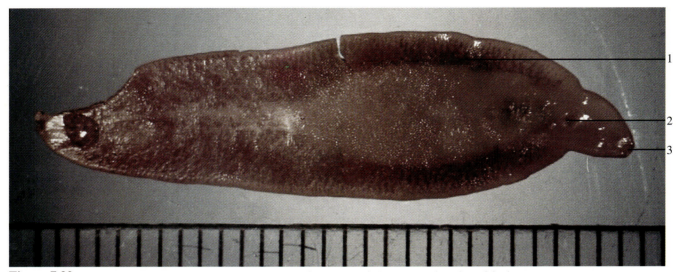

Figure 7.32   The cow liver fluke, *Fasciola magna,* is one of the largest flukes, measuring about 3 inches.
1. Yolk gland   2. Ventral sucker   3. Oral sucker

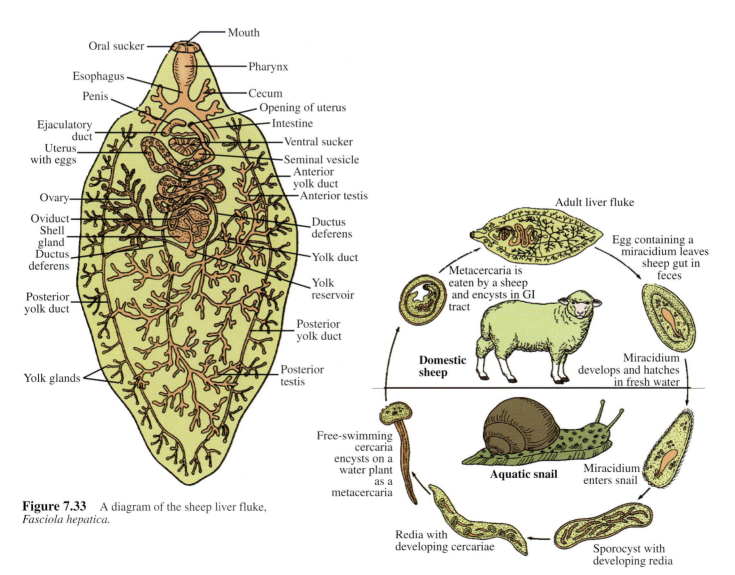

Figure 7.33   A diagram of the sheep liver fluke, *Fasciola hepatica.*

Figure 7.34   Life cycle of sheep liver fluke, *Fasciola hepatica.*

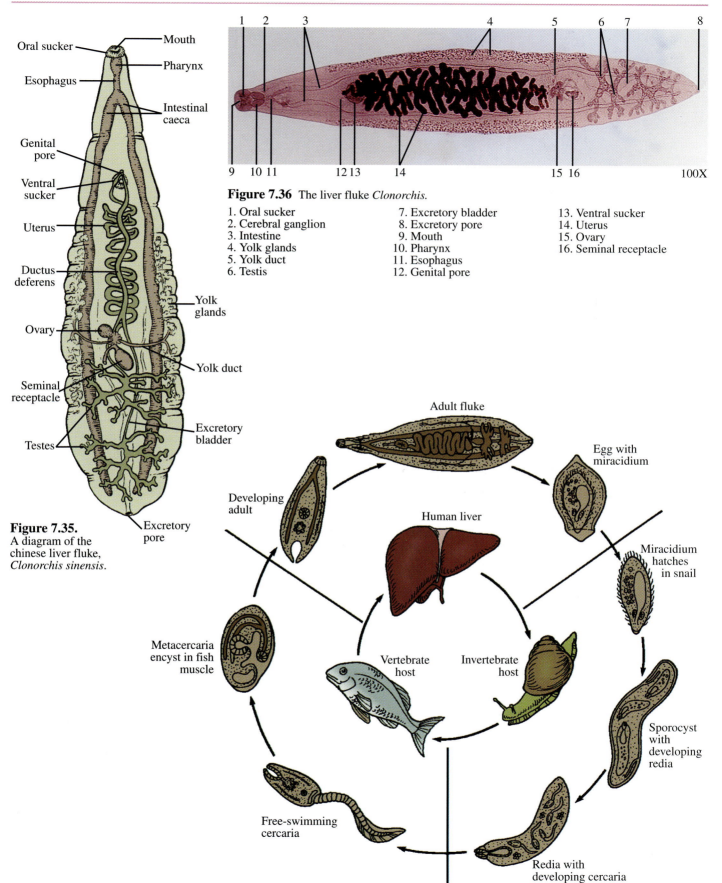

Oral sucker — — Mouth
— Pharynx
Esophagus —
— Intestinal caeca
Genital pore —
Ventral sucker —
Uterus —
Ductus deferens —
— Yolk glands
Ovary —
— Yolk duct
Seminal receptacle —
— Excretory bladder
Testes —
— Excretory pore

Figure 7.35.
A diagram of the chinese liver fluke, *Clonorchis sinensis*.

1   2   3   4   5   6   7   8

9   10 11   12 13   14   15 16   100X

Figure 7.36 The liver fluke *Clonorchis*.

1. Oral sucker
2. Cerebral ganglion
3. Intestine
4. Yolk glands
5. Yolk duct
6. Testis

7. Excretory bladder
8. Excretory pore
9. Mouth
10. Pharynx
11. Esophagus
12. Genital pore

13. Ventral sucker
14. Uterus
15. Ovary
16. Seminal receptacle

Adult fluke

Developing adult

Egg with miracidium

Human liver

Miracidium hatches in snail

Metacercaria encyst in fish muscle

Vertebrate host

Invertebrate host

Sporocyst with developing redia

Free-swimming cercaria

Redia with developing cercaria

Figure 7.37 Life cycle of the human liver fluke, *Clonorchis sinesis*.

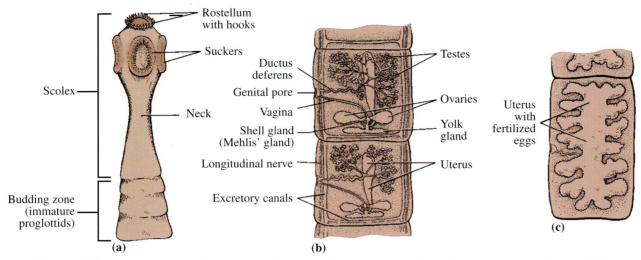

Figure 7.38 Diagrams of a parasitic tapeworm, *Taenia pisiformis.* (a) the anterior end, (b) mature proglottids, and (c) a ripe proglottid.

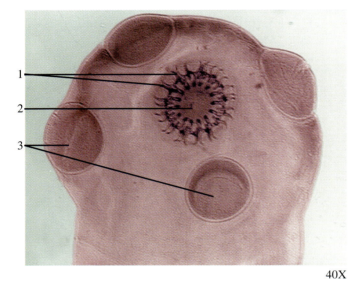

40X

Figure 7.39 Scolex of *Taenia pisiformis.*

1. Hooks
2. Rostellum
3. Suckers

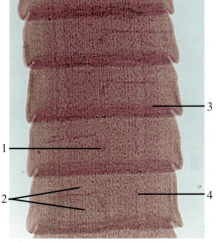

Figure 7.40 Immature   40X
proglottids of *Taenia pisiformis,*

1. Early ovary
2. Early testes
3. Excretory canal
4. Immature vagina and ductus deferens

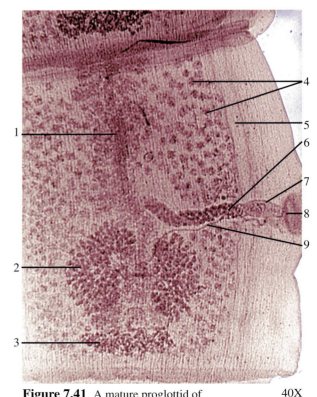

Figure 7.41 A mature proglottid of   40X
Taenia pisiformis.

1. Uterus	4. Testes	7. Cirrus
2. Ovary	5. Excretory canal	8. Genital pore
3. Yolk gland	6. Ductus deferens	9. Vagina

Nematoda — roundworms and nematodes

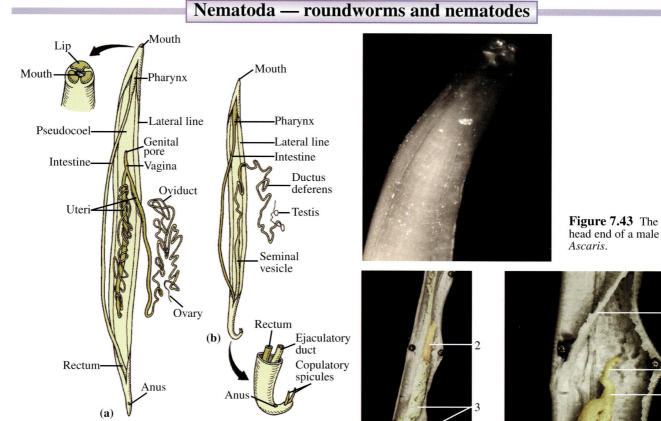

Figure 7.42   Diagrams of the internal anatomy of (a) a male and (b) a female *Ascaris*.

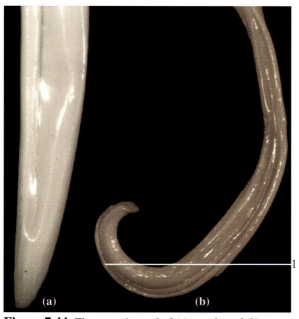

Figure 7.44   The posterior end of (a) a male and (b) a female *Ascaris*.

1. Ejaculatory duct

Figure 7.43   The head end of a male *Ascaris*.

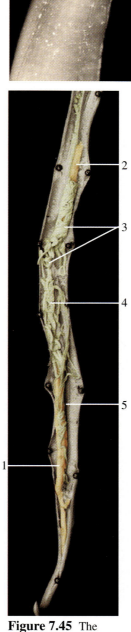

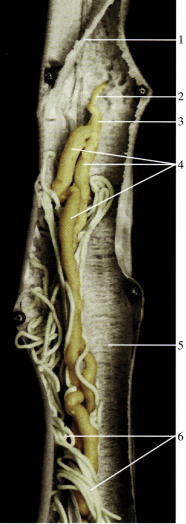

Figure 7.45   The internal anatomy of a male *Ascaris*.

1. Seminal vesicle
2. Intestine
3. Ductus deferens
4. Testes
5. Lateral line

Figure 7.46   The internal anatomy of a female *Ascaris*.

1. Intestine
2. Genital pore
3. Vagina
4. Uterus (Y-shaped)
5. Lateral line
6. Oviducts

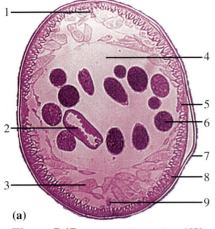

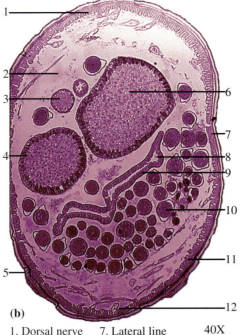

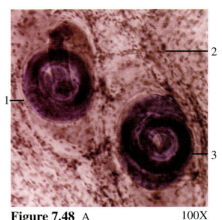

Figure 7.47 Cross sections of   40X
(a) a male and (b) a female *Ascaris.*

1. Dorsal nerve cord
2. Intestine
3. Longitudinal muscle cell body
4. Pseudocoel
5. Lateral line
6. Testis
7. Cuticle
8. Contractile sheath of muscle cell
9. Ventral nerve cord

(b)

1. Dorsal nerve	7. Lateral line 40X
2. Pseudocoel	8. Lumen of intestine
3. Oviduct	9. Intestine
4. Uterus	10. Ovary
5. Cuticle	11. Longitudinal muscles
6. Eggs	12. Ventral nerve cord

Figure 7.48 A   100X
photomicrograph of *Trichinella
spiralis* encysted in muscle.

1. Cyst
2. Muscle
3. Larva

Rotifera — rotifer

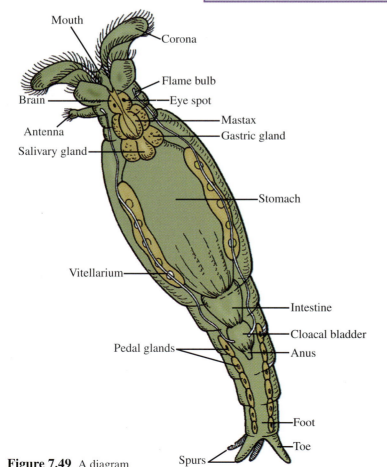

Mouth
Corona
Flame bulb
Brain
Eye spot
Antenna
Mastax
Gastric gland
Salivary gland
Stomach
Vitellarium
Intestine
Cloacal bladder
Pedal glands
Anus
Foot
Toe
Spurs

Figure 7.49 A diagram
of the rotifer, *Philodina.*

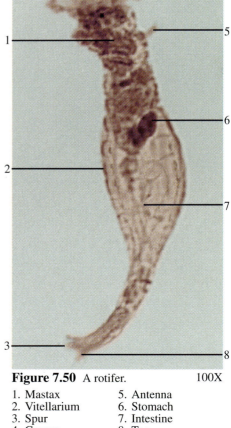

Figure 7.50 A rotifer.   100X

1. Mastax	5. Antenna
2. Vitellarium	6. Stomach
3. Spur	7. Intestine
4. Corona	8. Toe

Mollusca — Mollusks; clams, snails, and squids

Table 7.4 Some Representatives of the Phylum Mollusca

Classes and Representative Kinds	Characteristics
Polyplacophora — chitons	Marine; shell of 8 transverse plates; broad foot
Gastropoda — snails; slugs	Marine, freshwater, and terrestrial; coiled shell; prominent head with tentacles and eyes
Bivalvia — clams, oysters, mussels	Marine and freshwater; body compressed between two hinged shells; hatchet-shaped foot
Cephalopodes — squids, octopodes	Marine; excellent swimmers, predatory; foot separated into tentacles which may contain suckers; well-developed eyes

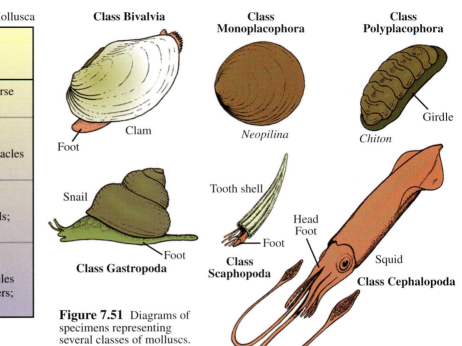

Figure 7.51 Diagrams of specimens representing several classes of molluscs.

Figure 7.52 Chitons are easily recognized by their eight dorsal plates.

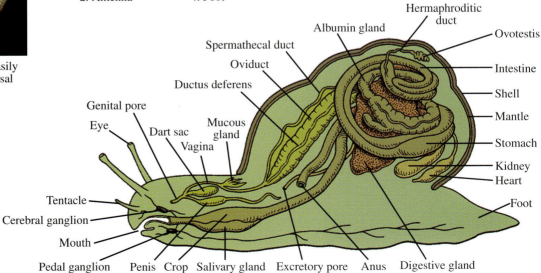

Figure 7.53 The locomotion of the slug, class Gastropoda, requires the production of mucus. Slugs differ from snails in that a shell is absent.

1. Head
2. Antenna
3. Mucus
4. Foot

Figure 7.54 A diagram of pulmonate snail anatomy.

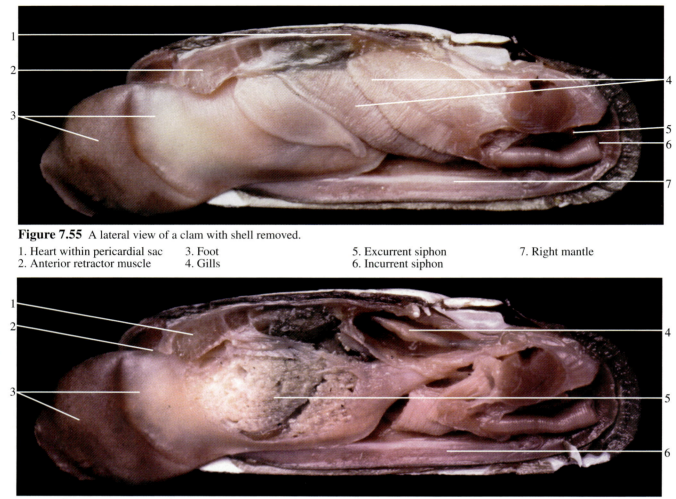

Figure 7.55 A lateral view of a clam with shell removed.

1. Heart within pericardial sac
2. Anterior retractor muscle
3. Foot
4. Gills
5. Excurrent siphon
6. Incurrent siphon
7. Right mantle

Figure 7.56 A lateral view of a clam with part of the visceral mass removed.

1. Stomach
2. Mouth
3. Foot
4. Heart
5. Gonad
6. Right mantle

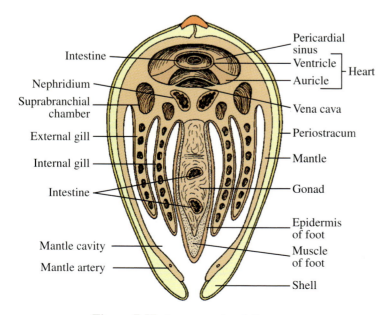

Figure 7.57 A cross-sectional diagram through the heart region of a freshwater clam.

Intestine
Nephridium
Suprabranchial chamber
External gill
Internal gill
Intestine
Mantle cavity
Mantle artery

Pericardial sinus
Ventricle ⎱ Heart
Auricle ⎰
Vena cava
Periostracum
Mantle
Gonad
Epidermis of foot
Muscle of foot
Shell

Figure 7.58 *Nautilus,* a cephalopod, has gas-filled chambers within its shell, as seen in this dissected specimen. These chambers regulate buoyancy.

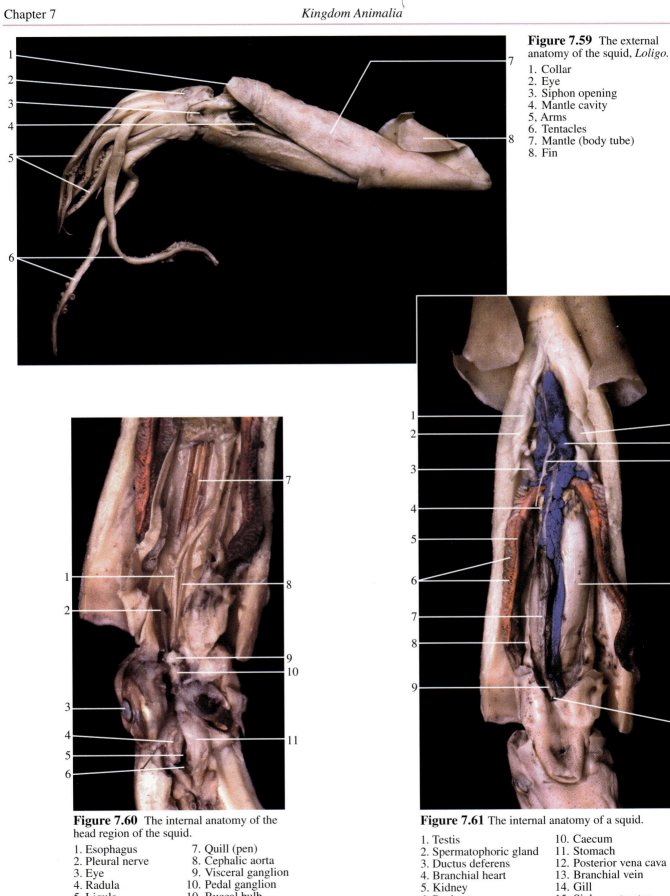

Figure 7.59 The external anatomy of the squid, *Loligo.*

1. Collar
2. Eye
3. Siphon opening
4. Mantle cavity
5. Arms
6. Tentacles
7. Mantle (body tube)
8. Fin

Figure 7.60 The internal anatomy of the head region of the squid.

1. Esophagus
2. Pleural nerve
3. Eye
4. Radula
5. Ligula
6. Mandible
7. Quill (pen)
8. Cephalic aorta
9. Visceral ganglion
10. Pedal ganglion
10. Buccal bulb

Figure 7.61 The internal anatomy of a squid.

1. Testis
2. Spermatophoric gland
3. Ductus deferens
4. Branchial heart
5. Kidney
6. Penis
7. Ink sac
8. Liver
9. Rectum
10. Caecum
11. Stomach
12. Posterior vena cava
13. Branchial vein
14. Gill
15. Siphon retractor muscle
16. Anterior vena cava
17. Anus

Annelida — segmented worms

Table 7.5 Some Representatives of the Phylum Annelida

Classes and Representative Kinds	Characteristics
Polychaeta — tubeworms, sandworms	Mostly marine; segments with parapodia
Oligochaeta — earthworms	Freshwater and burrowing terrestrial forms; small setae; poorly developed head
Hirudinea — leeches	Freshwater; most are blood-sucking parasites; lack setae; prominent muscular suckers

Figure 7.62 Tubeworms (also called sea feathers), *Sabellastarte indico,* shown here feeding.

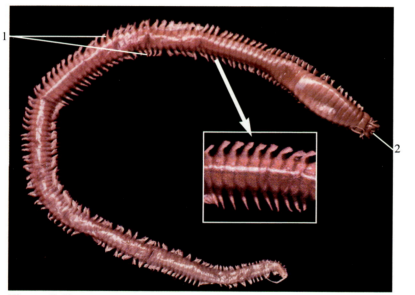

Figure 7.63 A dorsal view of the sandworm, *Neanthes.*
1. Parapodia　　　　　　2. Mouth

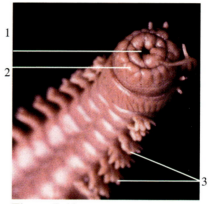

Figure 7.64 A view of the head of the sandworm, *Neanthes.*
1. Mouth
2. Everted pharynx
3. Prostomium

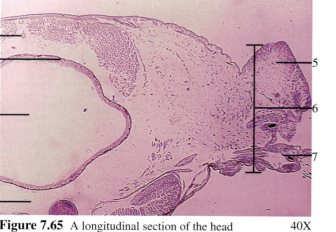

Figure 7.65 A longitudinal section of the head of the sandworm, *Neanthes.*　　　40X

1. Dorsal blood vessel　　4. Coelom　　　　　7. Neuropodium
2. Intestine　　　　　　　5. Notopodium
3. Lumen of gut　　　　　6. Parapodium

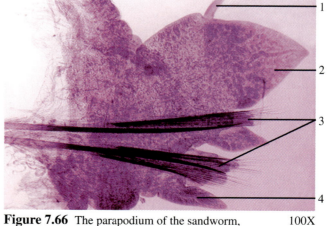

Figure 7.66 The parapodium of the sandworm, *Neanthes.*　　　100X

1. Dorsal cirrus　　　　3. Setae
2. Notopodium　　　　　4. Neuropodium

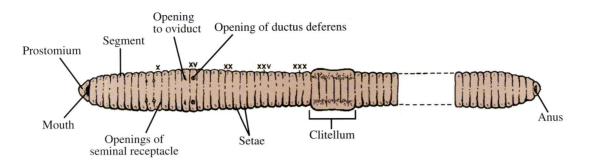

Figure 7.67 A ventral view diagram of the earthworm, *Lumbricus.*

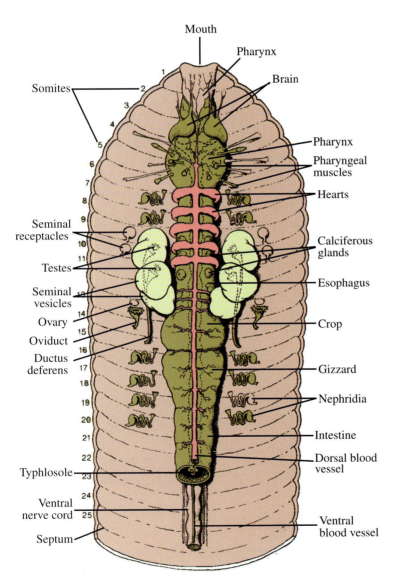

Figure 7.68 A diagram of the anterior end of the earthworm, *Lumbricus.*

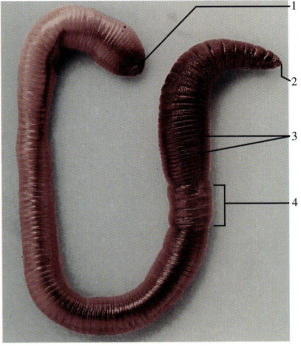

Figure 7.69 A dorsal view of an earthworm, *Lumbricus.*

1. Pygideum   3. Segments
2. Prostomium   4. Clitellum

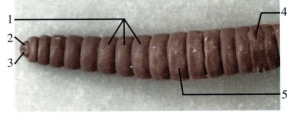

Figure 7.70 The anterior end of an earthworm, *Lumbricus.*

1. Setae   4. Opening of vas
2. Mouth   deferens
3. Prostomium   5. Segment 10

Figure 7.71 Earthworm cocoons (each line represents 1 mm).

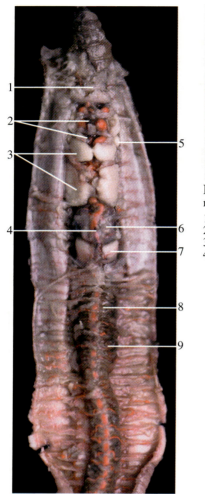

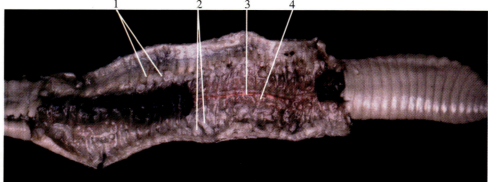

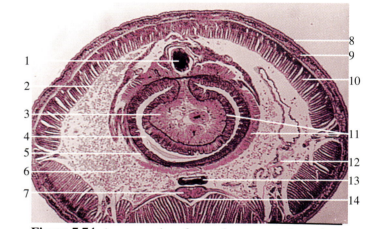

Figure 7.73 The internal anatomy of the posterior end of an earthworm with part of the intestine removed.

1. Septae
2. Nephridia
3. Dorsal blood vessel
4. Intestine

Figure 7.72 The internal anatomy of the anterior end of an earthworm, *Lumbricus.*

1. Pharynx
2. Hearts
3. Seminal vesicles
4. Dorsal blood vessel
5. Seminal recepticles
6. Crop
7. Gizzard
8. Intestine
9. Nephridia

Figure 7.74 A cross section of an earthworm posterior to the clitellum.

1. Dorsal blood vessel
2. Peritoneum
3. Typhlosole
4. Lumen of intestine
5. Intestine
6. Coelom
7. Ventral nerve cord
8. Epidermis
9. Circular muscles
10. Longitudinal muscles
11. Chloragogue cells
12. Nephridium
13. Ventral blood vessel
14. Subneural blood vessel

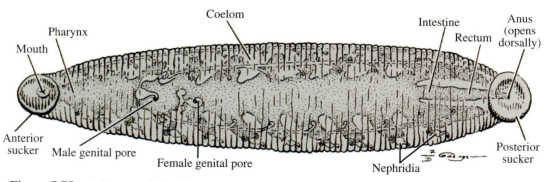

Figure 7.75 A diagram of a leech.

Figure 7.76 Leeches are more specialized than other annelids. They have lost their setae and developed suckers for attachment while sucking blood.

Arthropoda — arachnids, crustaceans, and insects

Table 7.6 Some Representatives of the Phylum Arthropoda

Classes and Representative Kinds	Characteristics
Merostomata — horseshoe crab	Cephalothorax and abdomen; specialized front appendages into chelicerae; lack antennae and mandibles
Arachnida — spiders, mites, ticks, scorpions	Cephalothorax and abdomen; four pairs of legs; book lungs or trachea; lack antennae and mandibles
Malacostraca — lobsters, crabs, shrimp	Cephalothorax and abdomen; two pair of antennae, pair of mandibles and maxillae; biramous appendages; gills
Insecta — beetles, butterflies, ants	Head, thorax, and abdomen; three pairs of legs; well-developed mouth parts; usually two pairs of wings; trachea
Chilopoda — centipedes	Head with segmented body; one pair of legs per segment; trachea; one pair of antennae
Diplopoda — millipedes	Head with segmented body; usually two pair of legs per segment; trachea

Figure 7.77 The trilobite, *Modicia typicalis,* is an extinct arthropod from the Cambrian and Ordovician periods.

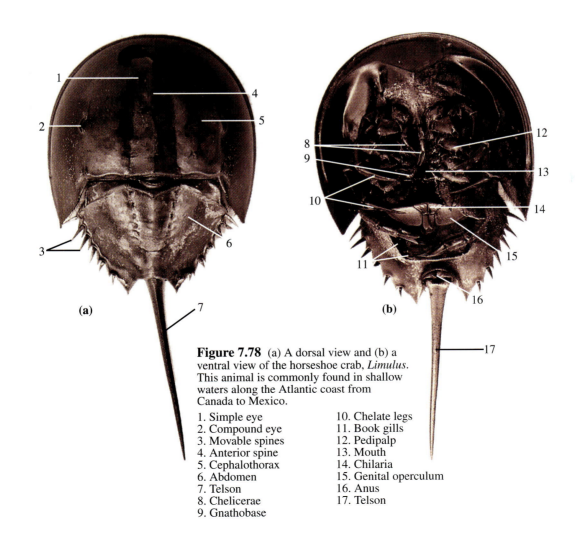

Figure 7.78 (a) A dorsal view and (b) a ventral view of the horseshoe crab, *Limulus.* This animal is commonly found in shallow waters along the Atlantic coast from Canada to Mexico.

1. Simple eye
2. Compound eye
3. Movable spines
4. Anterior spine
5. Cephalothorax
6. Abdomen
7. Telson
8. Chelicerae
9. Gnathobase
10. Chelate legs
11. Book gills
12. Pedipalp
13. Mouth
14. Chilaria
15. Genital operculum
16. Anus
17. Telson

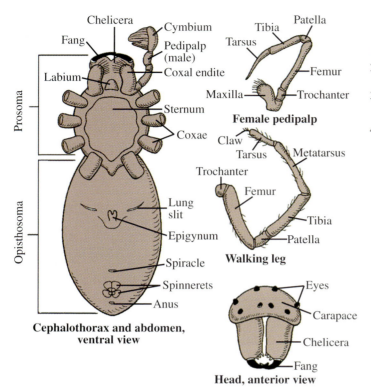

Cephalothorax and abdomen, ventral view

Female pedipalp

Walking leg

Head, anterior view

Figure 7.79 A diagram of the anatomy of the garden spider, *Argiope*.

Figure 7.80 The black widow spider, *Latrodectus mactans,* along with the brown recluse are the two spiders in the United States which can give severe or fatal bites.

1. Tibia	4. Tarsus	7. Prosoma
2. Femur	5. Patella	8. Trochanter
3. Metatarsus	6. Opisthosoma	

Figure 7.82 Tarantula, *Dugesiella,* in a defensive posture. The pautrons and the fangs comprise the chelicera.

1. Pautrons	3. Pedipalp	5. Prosoma
2. Opisthosoma	4. Fangs	

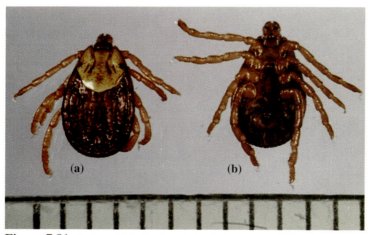

Figure 7.81 Ticks, within the family Ixodidae, are specialized parasitic arthropods. (a) A dorsal view and (b) a ventral view.

Figure 7.83 The scorpion, *Pandinus.* Scorpions are most commonly found in tropical and subtropical regions, but there are also several species found in temperate zones.

1. Stinging apparatus	4. Cephalothorax
2. Postabdomen (tail)	5. Preabdomen
3. Pedipalp	6. Walking legs

Figure 7.84 The water flea, *Daphnia*, is a common microscopic crustacean.

1. Heart
2. Midgut
3. Compound eye
4. Mouth
5. 2nd Antenna
6. Brood chamber
7. Hindgut
8. Abdominal seta
9. Anus
10. Thoracic appendages
11. Carapace
12. Setae

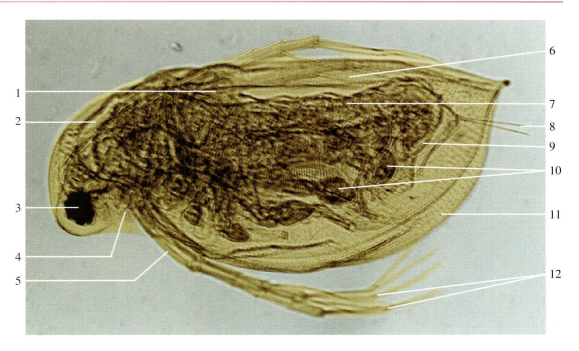

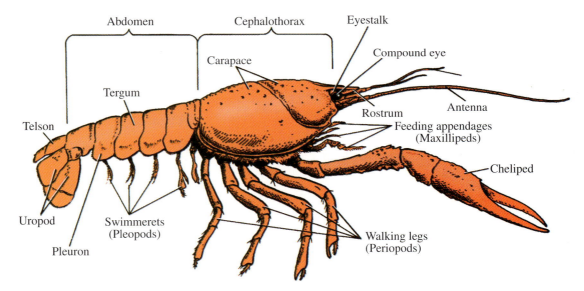

Figure 7.85 A diagram of the crayfish, *Cambarus*.

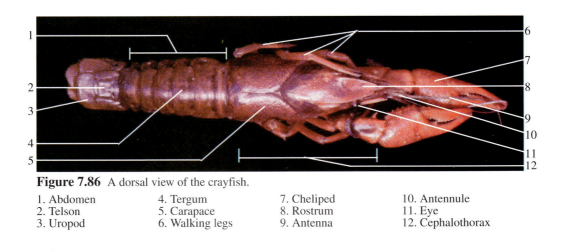

Figure 7.86 A dorsal view of the crayfish.

1. Abdomen	4. Tergum	7. Cheliped	10. Antennule
2. Telson	5. Carapace	8. Rostrum	11. Eye
3. Uropod	6. Walking legs	9. Antenna	12. Cephalothorax

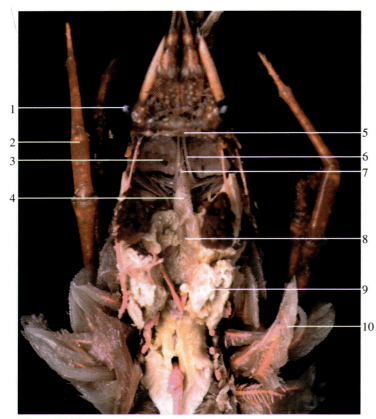

Figure 7.87 The oral region of the crayfish .

1. Eye
2. Walking leg
3. Green gland
4. Cardiac chamber of stomach
5. Brain
6. Circumesophageal connective (of ventral nerve cord)
7. Esophagus
8. Region of gastric mill
9. Digestive gland
10. Gill

Figure 7.89 A dorsal view of the anatomy of a crayfish.

1. Antenna
2. Compound eye
3. Brain
4. Circumesophageal connection (of ventral nerve)
5. Mandibular muscle
6. Digestive gland
7. Gills
8. Antennules
9. Walking legs
10. Green gland
11. Esophagus
12. Pyloric stomach
13. Testis
14. Ductus deferens
15. Aorta
16. Intestine

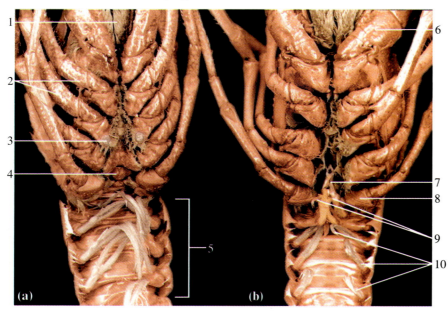

Figure 7.88 Ventral views of (a) a male and (b) a female crayfish. The first pair of swimmerets are greatly enlarged in the male for the depositing of sperm in the female's seminal receptacle.

1. Third maxilliped
2. Walking legs
3. Disc covering oviduct
4. Seminal receptacle
5. Abdomen
6. Chiliped
7. Copulatory swimmerets (pleopods)
8. Sperm ducts (genital pores)
9. Swimmerets (pleopods)

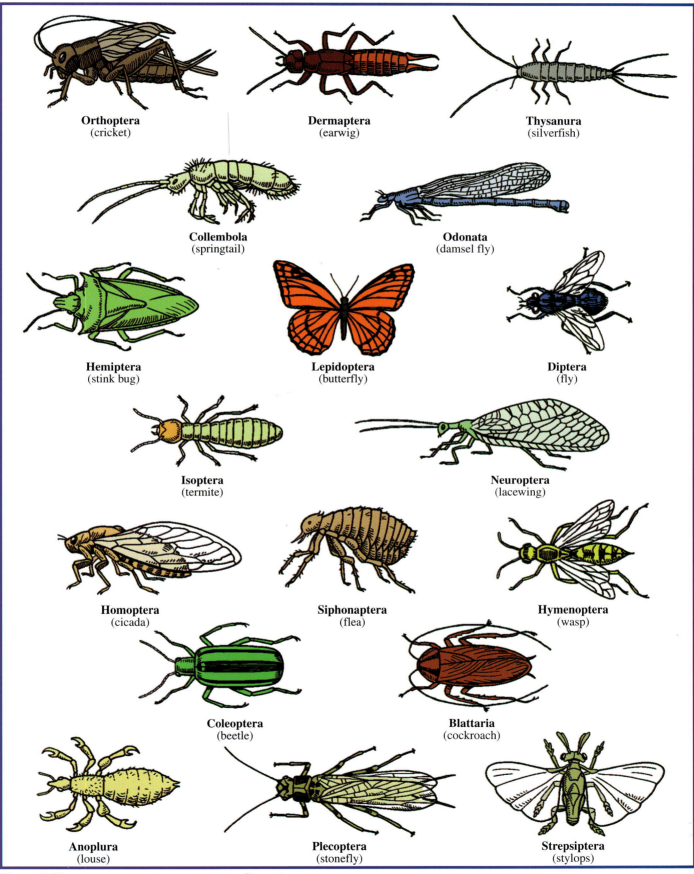

Figure 7.90 Representatives from some of the orders of insects.

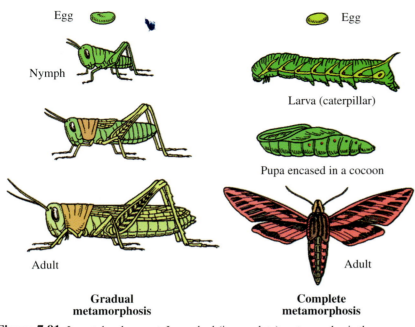

Gradual
metamorphosis

Complete
metamorphosis

Figure 7.91 Insect development. In gradual (incomplete) metamorphosis the young resemble the adults but they are smaller and have different body proportions. In complete metamorphosis, the larvae look different than the adult and generally have different food requirements.

Figure 7.92 The common grey cricket molting. All arthropods must periodically shed their exoskeleton in order to grow. This process is called molting, or ecdysis.

Figure 7.93 Developmental stages of the common honeybee include (a) larval stage, (b) pupa, and (c) adult.

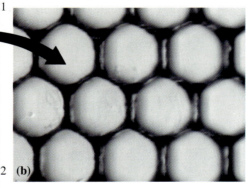

Figure 7.94 (a) A lateral view of the head of a butterfly. The most obvious structures on the head of a butterfly are the compound eyes and the curled tongue for siphoning nectar from flowers. (b) A magnified view of the compound eye.

1. Compound eye   2. Tongue

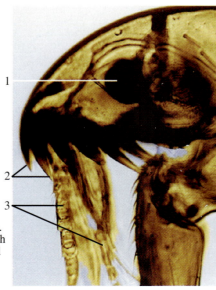

Figure 7.95 The mouthparts of the flea, *Ctenocephalides,* are specialized for parasitism. Notice the oral bristles beneath the mouth which aid the flea in penetrating between hairs to feed on the blood of mammals.

1. Eye   2. Oral bristles   3. Maxillary palps

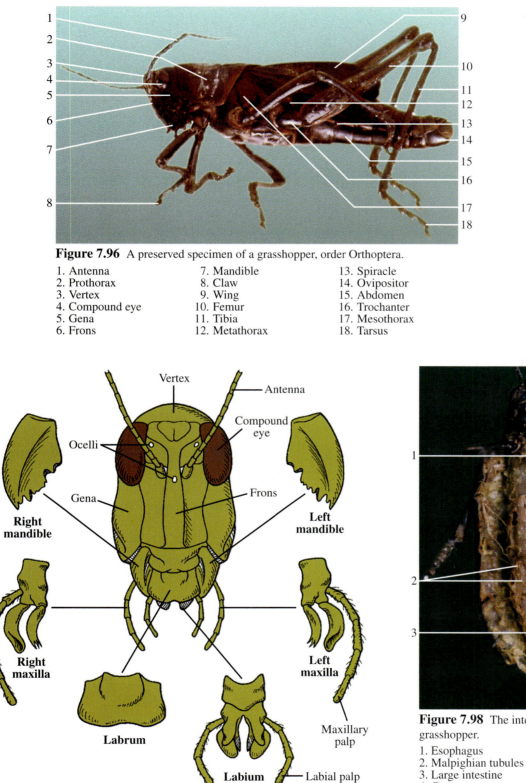

Figure 7.96 A preserved specimen of a grasshopper, order Orthoptera.

1. Antenna	7. Mandible	13. Spiracle
2. Prothorax	8. Claw	14. Ovipositor
3. Vertex	9. Wing	15. Abdomen
4. Compound eye	10. Femur	16. Trochanter
5. Gena	11. Tibia	17. Mesothorax
6. Frons	12. Metathorax	18. Tarsus

Figure 7.97 A diagram of the head and mouthparts of a grasshopper.

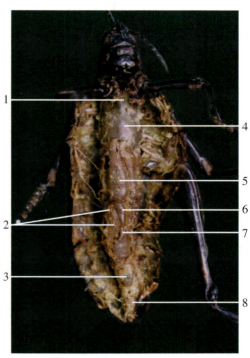

Figure 7.98 The internal anatomy of a grasshopper.

1. Esophagus	5. Gizzard
2. Malpighian tubules	6. Stomach
3. Large intestine	7. Intestine
4. Crop	8. Rectum

Enichnodermata — echinoderms, sea stars, sand dollars, sea cucumbers, and sea urchins

Table 7.7 Some Representatives of the Phylum Echinodermata

Classes and Representative Kinds	Characteristics
Asteroidea — sea stars (star fish)	Pentaradial symmetry; appendages arranged around a central disk containing the mouth;
Echinoidea — sea urchins, sand dollars	Disk-shaped with no arms; compact skeleton; movable spines; tube feet with suckers
Ophiuroidea — brittle stars	Pentaradial symmetry; appendages sharply marked off from central disk; tube feet without suckers
Holothuroidea — sea cucumbers	Cucumber-shaped with no arms; spines absent; tube feet with tentacles and suckers
Crinoidea — sea lilies and feather stars	Sessile during much of life cycle; calyx supported by elongated stalk

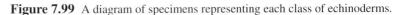

Sea urchin   Sand dollar
Class Echinoidea

Sea lily
Class Crinoidea

Sea stars
Class Asteroidea

Brittle star
Class Ophiuroidea

Sea cucumber
Class Holothuroidea

Figure 7.99 A diagram of specimens representing each class of echinoderms.

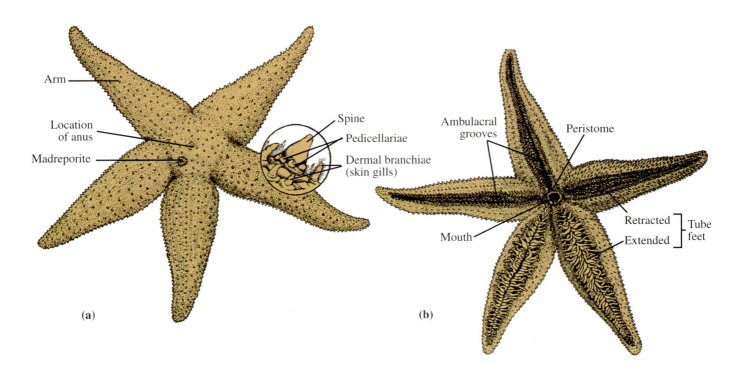

Arm

Location of anus

Madreporite

Spine

Pedicellariae

Dermal branchiae (skin gills)

Ambulacral grooves

Peristome

Mouth

Retracted ⎫ Tube
Extended ⎭ feet

(a)

(b)

Figure 7.100   A diagram of the external anatomy of the sea star (star fish), *Asterias*. (a) A dorsal (aboral) view and (b) a ventral (oral) view.

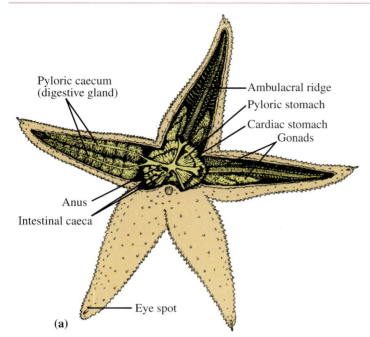

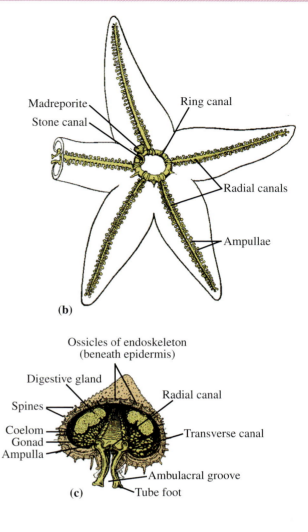

Figure 7.101   A diagram of the internal anatomy of the sea star. (a) The digestive and reproductive organs, (b) the water vascular system, and (c) a cross section through an arm.

Figure 7.102 A dorsal view of the internal anatomy of a sea star.

1. Ambulacral ridge
2. Ampullae
3. Pyloric duct
4. Gonad
5. Spines
6. Circular canal
7. Pyloric stomach

Figure 7.103 A magnified dorsal view of the internal anatomy of a sea star.

1. Ambulacral ridge
2. Madreporite
3. Stone canal
4. Ampullae
5. Polian vesicle
6. Pyloric caecum (digestive gland)
7. Gonad
8. Spines
9. Stomach
10. Spines

Figure 7.104 A ventral view of a sea star.

1. Tube feet
2. Pedicellariae
3. Peristome
4. Oral spines
5. Mouth
6. Ambulacral groove

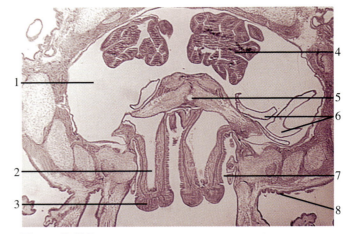

Figure 7.105 A cross section through the arm of a sea star.

1. Coelom
2. Tube foot
3. Sucker
4. Pyloric caecum
5. Radial canal
6. Gonad
7. Ambulacral groove
8. Pedicellaria

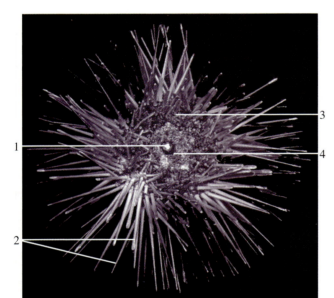

Figure 7.106 A dorsal (aboral) vew of the sea urchin, *Arbacia*.

1. Mouth   2. Spines   3. Pedicellaria   4. Peristome

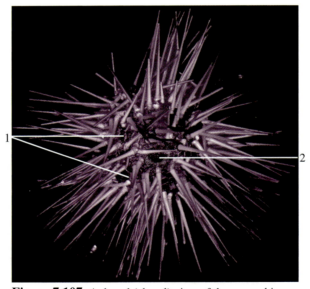

Figure 7.107 A dorsal (aboral) view of the sea urchin.

1. Ossicles   2. Madreporite

Figure 7.108 The internal anatomy of a sea urchin.

1. Madreporite
2. Gonad
3. Intestine
4. Aristotle's lantern
5. Mouth
6. Anus
7. Esophagus
8. Calcareous tooth
9. Stomach

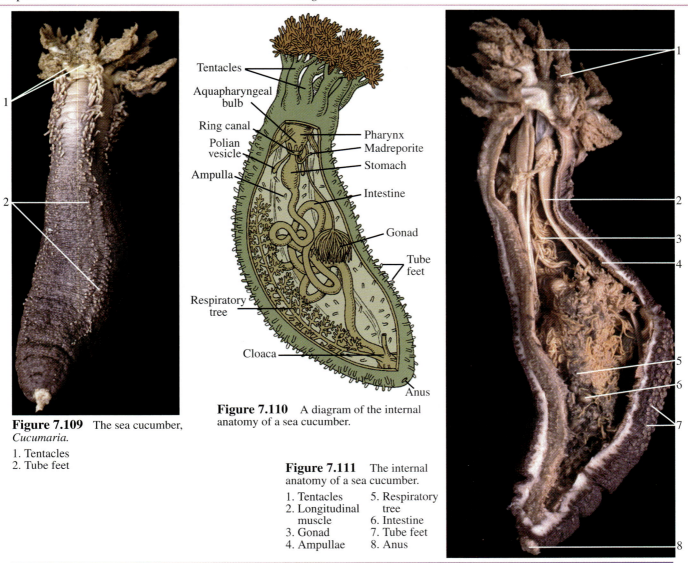

Figure 7.109   The sea cucumber, *Cucumaria.*

1. Tentacles
2. Tube feet

Figure 7.110   A diagram of the internal anatomy of a sea cucumber.

Tentacles
Aquapharyngeal bulb
Ring canal
Polian vesicle
Ampulla
Pharynx
Madreporite
Stomach
Intestine
Gonad
Tube feet
Respiratory tree
Cloaca
Anus

Figure 7.111   The internal anatomy of a sea cucumber.

1. Tentacles
2. Longitudinal muscle
3. Gonad
4. Ampullae
5. Respiratory tree
6. Intestine
7. Tube feet
8. Anus

Chordata — Amphioxus; amphibians, fishes, reptiles, birds, and mammals

Table 7.8 Some Representatives of the Phylum Chordate.

Subphyla and Representative Kinds	Characteristics
Urochordata: tunicates	Marine, larvae are free-swimming and have notochord, gill slits, and dorsal hollow nerve cord; adults are sessile (attached), filter-feeders, saclike animals
Cephalochordata: lancelets (*Amphioxus*)	Marine, segmented, elongated body with notochord extending the length of the body; cilia surrounding the mouth for obtaining food
Vertebrata: agnathans (lampreys and hagfishes), fishes (cartilaginous and bony), amphibians, reptiles, birds, mammals	Aquatic and terrestrial forms; distinct head and trunk supported by a series of cartilaginous or bony vertebrae in the adult; closed circulatory system and ventral heart; well-developed brain and sensory organs

Table 7.9 Some Representatives of the Subphylum Vertebrate.

Class and Representative Kinds	Characteristics
Agnatha — hagfish, lamprey	Eel-like and aquatic; sucking mouth (some parasitic); lack paired appendages
Chondrichthyes — sharks, rays, skates	Cartilaginous skeleton; placoid scales; spiracle; spiral valve in digestive tract
Osteichthyes — bony fishes	Gills covered by bony operculum; most have swim bladder
Amphibia — salamanders, frogs, toads	Larvae have gills and adults have lungs; scaleless skin; incomplete double circulation
Reptilia — turtles, snakes, lizards	Amniotic egg; epidermal scales; three- or four-chambered heart; lungs
Aves — birds	Homeothermous (warm-blooded); feathers; toothless; air sacs; four-chambered heart with right aortic arch
Mammalia — mammals	Homeothermous; hair; mammary glands; seven cervical vertebrae; muscular diaphragm; three auditory ossicles; four-chambered heart with left aortic arch

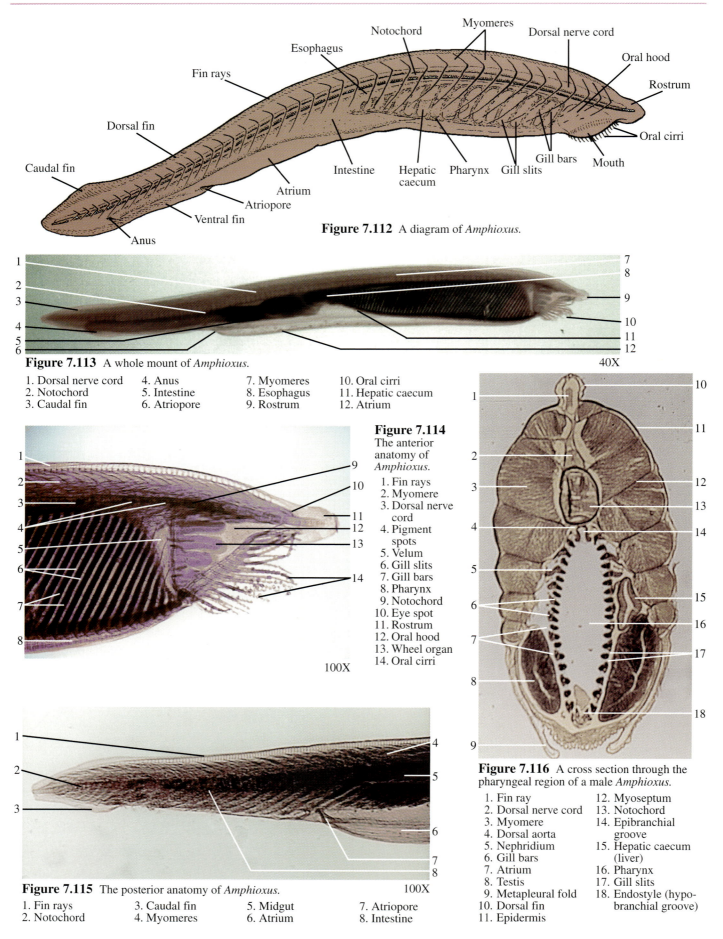

Figure 7.112 A diagram of *Amphioxus*.

Figure 7.113 A whole mount of *Amphioxus*.   40X

1. Dorsal nerve cord	4. Anus	7. Myomeres	10. Oral cirri
2. Notochord	5. Intestine	8. Esophagus	11. Hepatic caecum
3. Caudal fin	6. Atriopore	9. Rostrum	12. Atrium

Figure 7.114
The anterior
anatomy of
Amphioxus.

1. Fin rays
2. Myomere
3. Dorsal nerve
   cord
4. Pigment
   spots
5. Velum
6. Gill slits
7. Gill bars
8. Pharynx
9. Notochord
10. Eye spot
11. Rostrum
12. Oral hood
13. Wheel organ
14. Oral cirri

100X

Figure 7.116 A cross section through the
pharyngeal region of a male *Amphioxus*.

1. Fin ray	12. Myoseptum
2. Dorsal nerve cord	13. Notochord
3. Myomere	14. Epibranchial
4. Dorsal aorta	groove
5. Nephridium	15. Hepatic caecum
6. Gill bars	(liver)
7. Atrium	16. Pharynx
8. Testis	17. Gill slits
9. Metapleural fold	18. Endostyle (hypo-
10. Dorsal fin	branchial groove)
11. Epidermis	

Figure 7.115 The posterior anatomy of *Amphioxus*.   100X

1. Fin rays	3. Caudal fin	5. Midgut	7. Atriopore
2. Notochord	4. Myomeres	6. Atrium	8. Intestine

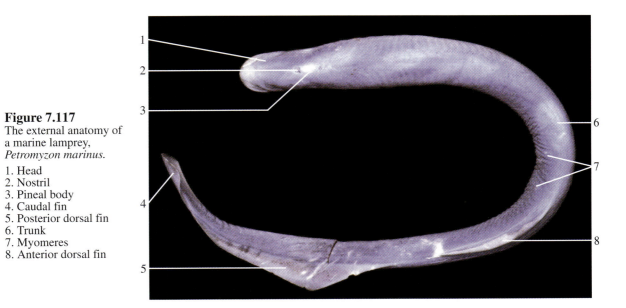

Figure 7.117
The external anatomy of
a marine lamprey,
Petromyzon marinus.

1. Head
2. Nostril
3. Pineal body
4. Caudal fin
5. Posterior dorsal fin
6. Trunk
7. Myomeres
8. Anterior dorsal fin

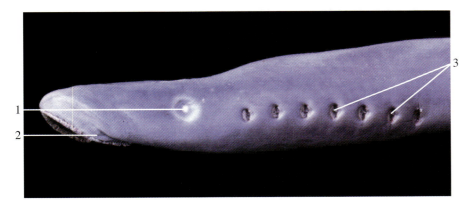

Figure 7.118
The anterior anatomy of a
marine lamprey.

1. Eye
2. Buccal funnel
3. External gill slits

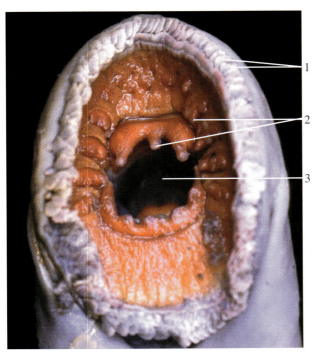

Figure 7.119
The oral region of a
marine lamprey.

1. Buccal papillae
2. Horny teeth
3. Mouth

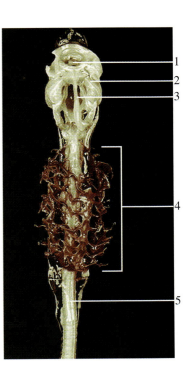

Figure 7.120
A cartilaginous skeleton of
a marine lamprey.

1. Buccal cavity
2. Skull
3. Tongue support
4. Branchial basket
5. Notochord

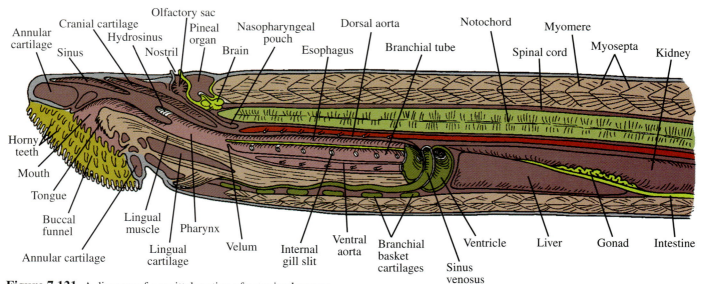

Figure 7.121 A diagram of a sagittal section of a marine lamprey.

Figure 7.122 A sagittal section through the anterior region of a lamprey.

1. Pineal organ	5. Mouth	9. Esophagus	13. Dorsal aorta
2. Nostril	6. Annular cartilage	10. Myomeres	14. Ventricle
3. Brain	7. Lingual cartilage	11. Spinal cord	15. Sinus venosus
4. Annular cartilage	8. Pharynx	12. Notochord	16. Internal gill slits

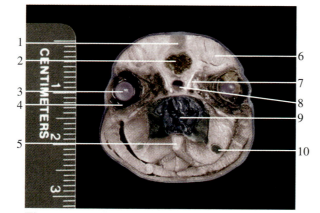

Figure 7.123 A cross section through the eyes of a lamprey.

1. Pineal organ	6. Myomere
2. Brain	7. Cranial cartilage
3. Lens of eye	8. Nasopharyngeal pouch
4. Retina of eye	9. Pharynx
5. Lingual cartilage	10. Pharyngeal gland

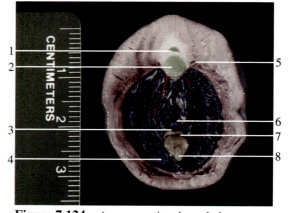

Figure 7.124 A cross section through the branchial tube anterior to the fourth gill pouch of a lamprey. The ventral aorta is paired at this location.

1. Spinal cord	5. Anterior cardinal vein
2. Notochord	6. Esophagus
3. Branchial tube	7. Ventral aorta
4. Gill pouch	8. Lingual muscle

Figure 7.125
A dorsal view of a
skate, *Raja.*

1. Rostrum
2. Eyes
3. Pectoral fin
4. Tail
5. Pelvic fin

Figure 7.126
A lateral view of
the leopard shark,
*Triakis
semifasciata.*

1. Spiracle
2. Lateral line
3. Dorsal fins
4. Caudal fin
   (heterocercal tail)
5. Eye
6. Gill slits
7. Pectoral fin
8. Pelvic fin

Dorsal fin

Lateral line

Caudal fin

Adipose fin

Eye

Nostril

Mandible

Maxilla

Opercle

Pectoral fin

Pelvic fin

Anal fin

Illustration courtesy of
Joseph R. Tomelleri

Figure 7.127 External structures of a rainbow trout, *Salmo gairdneri.*

Figure 7.128 The brown salamander, *Ambystoma gracile.*
This amphibian lives in humid sites, often beneath debris along
stream banks.

(a)

Figure 7.129 An
adult Indonesian
giant tree frog,
Litoria infrafrenata.
(a) The frog is
crouched on a
person's fingers.
(b) The suction
cups can be seen in
a ventral view.

(b)

Figure 7.130 A photograph showing hatching king snakes.
Most snakes are oviparous, meaning they lay eggs such as these.
Some snakes, including all American pit vipers, are
ovoviviparous, giving birth to well-developed young.

Figure 7.131 A spurred tortoise, *Geochelone sulcata.*

Figure 7.132 A veiled chameleon, *Chamaeleo calyptratus.*
Chameleons are best known for their ability to change colors
according to their surroundings.

Figure 7.133 Representatives from some of the orders of birds.

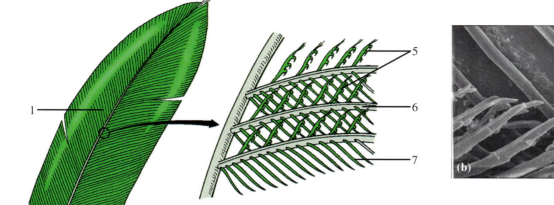

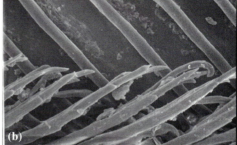

Figure 7.134 (a) The structure of a contour (pluma) feather. (b) The barbules and hooklets are shown in a photomicrograph.

1. Vane
2. Rachis
3. Calamus
4. Shaft
5. Hooklets
6. Barb
7. Barbule

Figure 7.135 Birds are adapted to occupy specific ecological niches. (a) A flamingo, *Phoenicopterus chilensis,* (b) an osprey, *Pandion haliaetus,* (c) a Northern flicker, *Colaptes auratus,* and (d) a scarlet macaw, *Ara macao.*

Figure 7.136 Representatives from the orders of mammals.

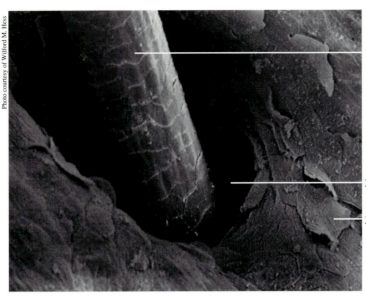

Photo courtesy of Wilford M. Hess

Figure 7.137 An electron micrograph of a hair emerging from a hair follicle.

1. Shaft of hair (note the scale-like pattern)
2. Hair follicle
3. Epithelial cell from stratum cornium

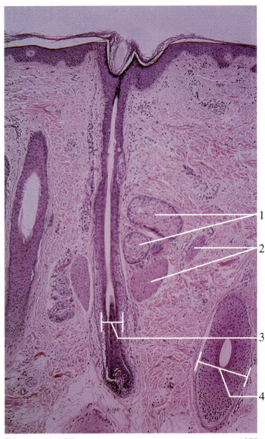

Figure 7.138 A photomicrograph of a hair and sebaceous gland.   40X

1. Sebaceous gland
2. Arrector pili muscle
3. Hair follicle
4. Hair follicle (oblique cut)

Figure 7.139 Museum prepared mammal skins are important in mammalian taxonomy.

Human Biology

8

The study of human biology requires learning about the anatomy and physiology of the human body. *Human anatomy* is the scientific discipline that investigates the structure of the body, and *human physiology* is the scientific discipline that investigates how body structures function. The purpose of this chapter is to present a visual overview of the principal anatomical structures of the human body.

Since both the *skeletal system* and the *muscular system* are both concerned with body movement, they are frequently discussed together as the *skeletomuscular system*. In a functional sense, the internal framework, or bones of the skeleton, support and provide movement at the joints where the muscles attached to the bones produce their actions as they are stimulated to contract.

The *nervous system* is anatomically divided into the *central nervous system* (CNS), which includes the *brain* and *spinal cord*, and the *peripheral nervous system* (PNS), which includes the *cranial nerves*, arising from the brain, and the *spinal nerves*, arising from the spinal cord. The *autonomic nervous system* (ANS) is a functional division of the nervous system devoted to regulation of involuntary activities of the body. The brain and spinal cord are the centers for integration and coordination of information. *Nerves*, composed of *neurons*, convey nerve impulse to and from the brain or spinal cord. Sensory organs, such as the eyes and ears, respond to environmental stimuli and convey sensations to the CNS. The nervous system functions with the *endocrine system* in coordinating body activities.

The *cardiovascular system* consists of the heart, vessels (both blood and lymphatic vessels), blood, and the tissues that produce the blood. The 4-chambered human heart is enclosed in a *pericardial sac* within the thoracic cavity. *Arteries* and *arterioles* transport blood away from the heart, *capillaries* permeate the tissues and are the functional units for product exchange with the cells, and *venules* and *veins* transport blood toward the heart. *Lymphatic vessels* return interstitial fluid back to the circulatory system after first passing it through *lymph nodes* for cleansing. Blood cells are produced in the bone marrow. Blood cells are broken down in the liver after they are old and worn.

The *respiratory system* consists of the conducting division that transports air to and from the respiratory division within the lungs. The *pulmonary alveoli* of the lungs are in contact with the capillaries of the cardiovascular system and are the sites for exchange of respiratory gases.

The *digestive system* consists of a *gastrointestinal tract* (GI tract) and *accessory digestive organs*. Food traveling through the GI tract is processed such that it is suitable for absorption through the intestinal wall into the blood. The *liver*

and *pancreas* are the principal digestive organs that process nutrients for body utilization.

Because of commonality of prenatal development and dual functions of some of the organs, the *urinary system* and *reproductive system* may be considered together as the *urogenital system*. The urinary system, consisting of the *kidneys*, *ureters*, *urinary bladder*, and *urethra*, extracts and processes metabolic wastes from the blood in the form of urine. The male and female reproductive systems produce regulatory hormones and *gametes* (sperm and ova, respectively) within the *gonads* (testes and ovaries). Sexual reproduction is the mechanism for producing offspring that have traits from both parents. The process of prenatal development is made possible by the formation of *extraembryonic membranes* (placenta, umbilical cord, allantois, amnion, and yolk sac) inside the uterus of the mother.

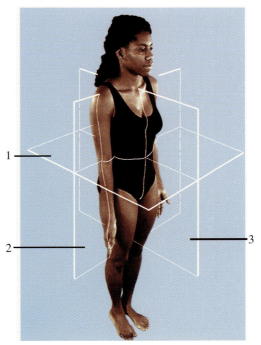

Figure 8.1 The planes of reference in a person while standing in anatomical position. The anatomical position provides a basis of reference for describing the relationship of one body part to another. In the anatomical position, the person is standing, the feet are parallel, the eyes are directed forward, and the arms are to the sides with the palms turned forward and the fingers are pointed straight down.

1. Transverse plane
   (cross-sectional plane)
2. Coronal plane
   (frontal plane)
3. Midsagittal plane
   (median plane)

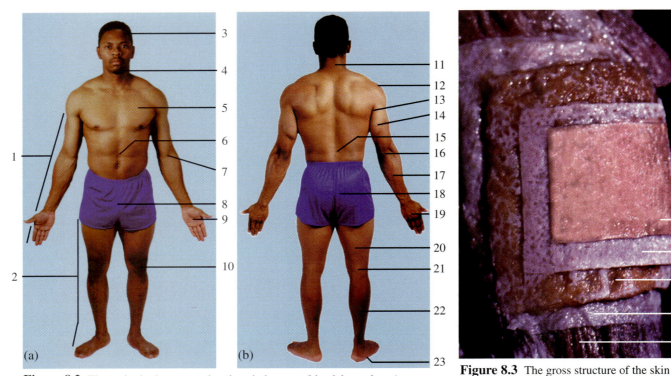

Figure 8.2 The major body parts and regions in humans (bipedal vertebrate). (a) An anterior view and (b) a posterior view.

1. Upper extremity	11. Cervical region	20. Thigh
2. Lower extremity	12. Shoulder	21. Popliteal fossa
3. Head	13. Axilla (armpit)	22. Calf
4. Neck, anterior aspect	14. Brachium (upper arm)	23. Plantar surface
5. Thorax (chest)	15. Lumbar region	(sole)
6. Abdomen	16. Elbow	
7. Cubital fossa	17. Antebrachium	
8. Pubic region	(forearm)	
9. Palmar region (palm)	18. Gluteal region	
10. Patellar region	(buttock)	
(patella)	19. Dorsum of hand	

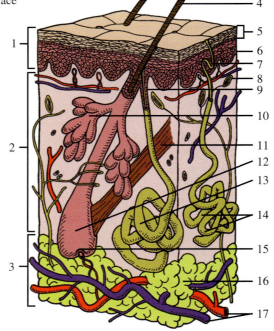

Figure 8.3 The gross structure of the skin and underlying fascia.

1. Epidermis   4. Fascia
2. Dermis   5. Muscle
3. Hypodermis

Figure 8.4 The epidermis and dermis of the skin.   75X

1. Stratum corneum	4. Stratum spinosum
2. Stratum lucidum	5. Stratum basale
3. Stratum granulosum	6. Dermis

Figure 8.5 The skin and associated structures.

1. Epidermis	10. Sebaceous gland
2. Dermis	11. Arrector pili muscle
3. Hypodermis	12. Hair follicle
4. Shaft of hair	13. Apocrine sweat gland
5. Stratum corneum	14. Eccrine sweat gland
6. Stratum basale	15. Bulb of hair
7. Sweat duct	16. Adipose tissue
8. Sensory receptor	17. Cutaneous blood vessels
9. Sweat duct	

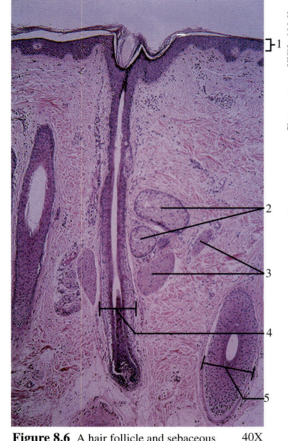

Figure 8.6 A hair follicle and sebaceous gland. 40X

1. Epidermis
2. Sebaceous glands
3. Arrector pili muscle
4. Hair follicle
5. Hair follicle (oblique cut)

Photo courtesy of Wilford M. Hess

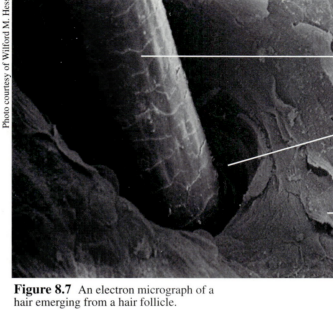

Figure 8.7 An electron micrograph of a hair emerging from a hair follicle.

1. Shaft of hair (note the scale–like pattern)
2. Hair follicle
3. Epithelial cell from stratum corneum

Skeletomusculature system

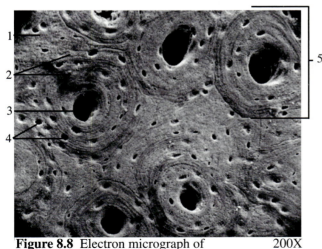

Figure 8.8 Electron micrograph of bone tissue. 200X

1. Interstitial lamellae
2. Lamellae
3. Central canal (haversian canal)
4. Lacunae
5. Osteon (haversian system)

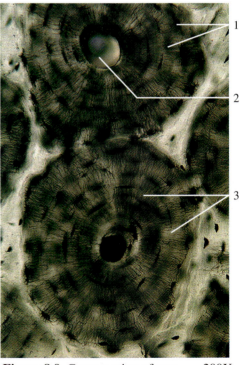

Figure 8.9 Cross section of two osteons. 200X

1. Lacunae
2. Central (haversian) canal
3. Lamellae

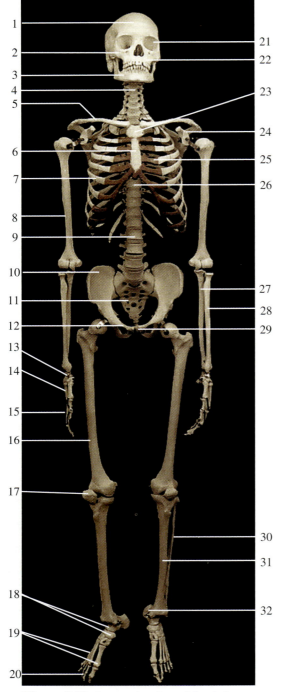

Figure 8.10 An anterior view of the skeleton.

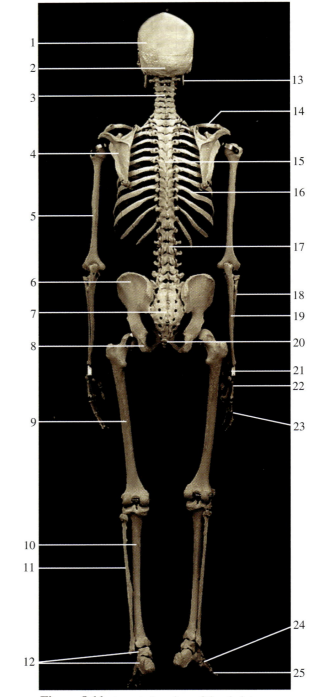

Figure 8.11 A posterior view of the skeleton.

1. Frontal bone	14. Metacarpal bones	27. Ulna
2. Zygomatic bone	15. Phalanges	28. Radius
3. Mandible	16. Femur	29. Symphysis pubis
4. Cervical vertebra	17. Patella	30. Fibula
5. Clavicle	18. Tarsal bones	31. Tibia
6. Body of sternum	19. Metatarsal bones	32. Calcaneus
7. Rib	20. Phalanges	
8. Humerus	21. Orbit	
9. Lumbar vertebra	22. Maxilla	
10. Ilium	23. Manubrium	
11. Sacrum	24. Scapula	
12. Pubis	25. Costal cartilage	
13. Carpal bones	26. Thoracic vertebra	

1. Parietal bone	11. Fibula	21. Carpal bones
2. Occipital bone	12. Tarsal bones	22. Metacarpal bones
3. Cervical vertebra	13. Mandible	23. Phalanges
4. Scapula	14. Clavicle	24. Metatarsal bones
5. Humerus	15. Thoracic vertebra	25. Phalanges
6. Ilium	16. Rib	
7. Sacrum	17. Lumbar vertebra	
8. Ischium	18. Radius	
9. Femur	19. Ulna	
10. Tibia	20. Coccyx	

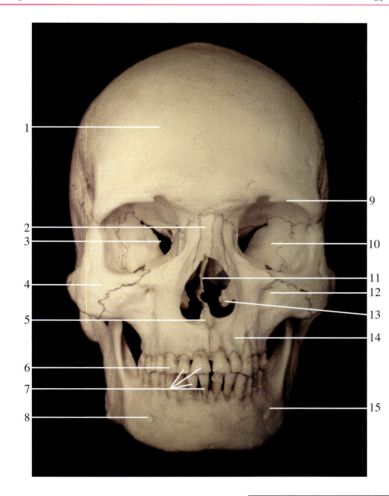

Figure 8.12 An anterior view of the skull.

1. Frontal bone
2. Nasal bone
3. Superior orbital fissure
4. Zygomatic bone
5. Vomer
6. Canine
7. Incisors
8. Mental foramen
9. Supraorbital margin
10. Sphenoid bone
11. Perpendicular plate of ethmoid bone
12. Infraorbital foramen
13. Inferior nasal concha
14. Maxilla
15. Mandible

Figure 8.13 A lateral view of the skull.

1. Coronal suture
2. Frontal bone
3. Lacrimal bone
4. Nasal bone
5. Zygomatic bone
6. Maxilla
7. Premolars
8. Molars
9. Mandible
10. Parietal bone
11. Squamosal suture
12. Temporal bone
13. Lambdoidal suture
14. External acoustic meatus
15. Occipital bone
16. Condylar process of mandible
17. Mandibular notch
18. Mastoid process of temporal bone
19. Coronoid process of mandible
20. Angle of mandible

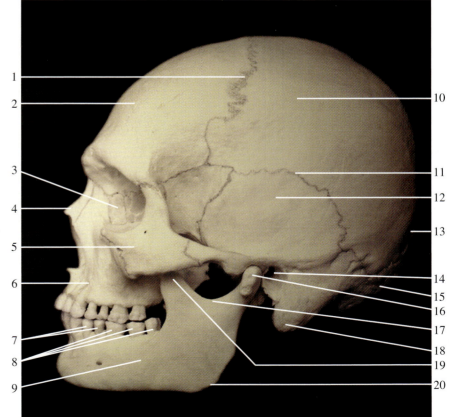

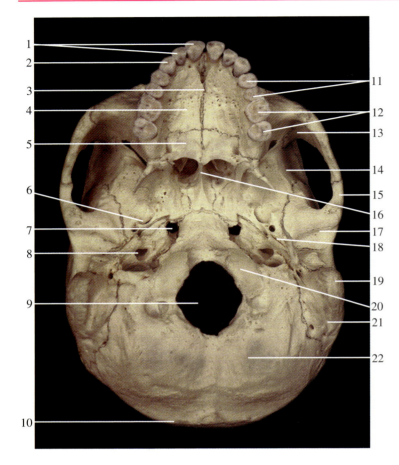

Figure 8.14 An inferior view of the skull.

1. Incisors
2. Canine
3. Intermaxillary suture
4. Maxilla
5. Palatine bone
6. Foramen ovale
7. Foramen lacerum
8. Carotid canal
9. Foramen magnum
10. Superior nuchal line
11. Premolars
12. Molars
13. Zygomatic bone
14. Sphenoid bone
15. Zygomatic arch
16. Vomer
17. Mandibular fossa
18. Styloid process of temporal bone
19. Mastoid process of temporal bone
20. Occipital condyle
21. Temporal bone
22. Occipital bone

Figure 8.15 A sagittal view of the skull.

1. Frontal bone
2. Frontal sinus
3. Crista galli of ethmoid bone
4. Cribriform plate of ethmoid bone
5. Nasal bone
6. Nasal concha
7. Maxilla
8. Mandible
9. Parietal bone
10. Occipital bone
11. Internal acoustic meatus
12. Sella turcica
13. Hypoglossal canal
14. Sphenoidal sinus
15. Styloid process of temporal bone
16. Vomer

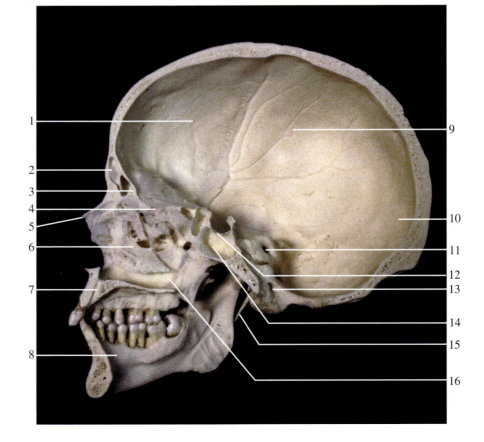

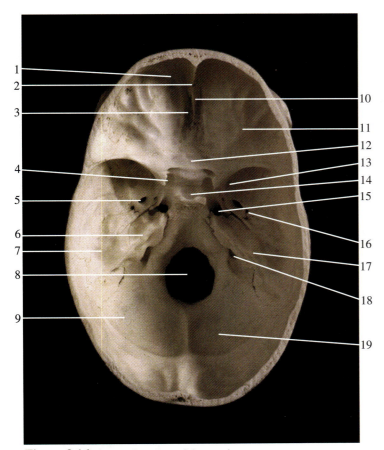

Figure 8.16 A superior view of the cranium.

1. Frontal bone
2. Foramen cecum
3. Cribriform plate of
   ethmoid bone
4. Optic canal
5. Foramen ovale
6. Petrous part of
   temporal bone
7. Temporal bone
8. Foramen magnum

9. Occipital bone
10. Crista galli of ethmoid bone
11. Anterior cranial fossa
12. Sphenoid bone
13. Foramen rotundum
14. Sella turcica of sphenoid bone
15. Foramen lacerum
16. Foramen spinosum
17. Internal acoustic meatus
18. Jugular foramen

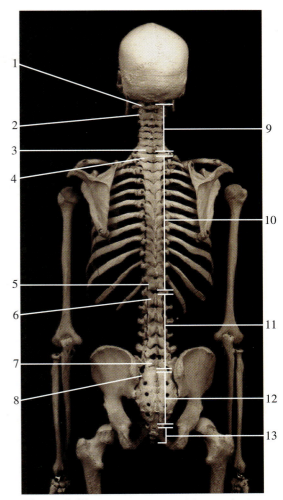

Figure 8.17 A posterior view of the vertebral
column.

1. Atlas
2. Axis
3. Seventh cervical
   vertebra
4. First thoracic vertebra
5. Twelfth thoracic
   vertebra
6. First lumbar vertebra

7. Fifth lumbar vertebra
8. Sacroiliac joint
9. Cervical vertebrae
10. Thoracic vertebrae
11. Lumbar vertebrae
12. Sacrum
13. Coccyx

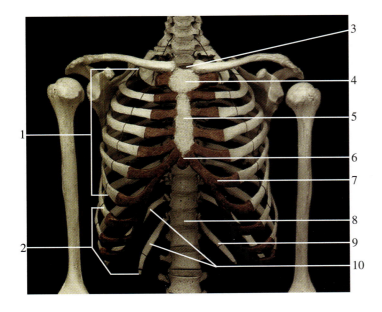

Figure 8.18 An anterior view of the rib cage.

1. True ribs (upper seven pairs of ribs)
2. False ribs (lower five pairs of ribs)
3. Jugular notch
4. Manubrium
5. Body of sternum
6. Xiphoid process
7. Costal cartilage
8. Twelfth thoracic vertebra
9. Twelfth rib
10. Floating ribs (lower two pairs of false ribs)

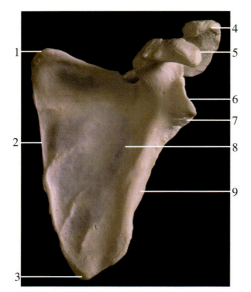

Figure 8.19 An anterior view of the left scapula.

1. Superior angle
2. Medial (vertebral) border
3. Inferior angle
4. Acromion
5. Coracoid process
6. Glenoid fossa
7. Infraglenoid tubercle
8. Subscapular fossa
9. Lateral (axillary) border

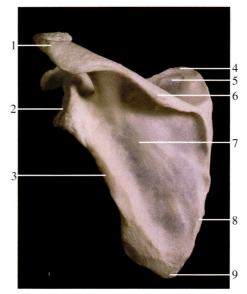

Figure 8.20 A posterior view of the left scapula.

1. Acromion
2. Glenoid fossa
3. Lateral (axillary) border
4. Superior angle
5. Supraspinous fossa
6. Spine
7. Infraspinous fossa
8. Medial (vertebral) border
9. Inferior angle

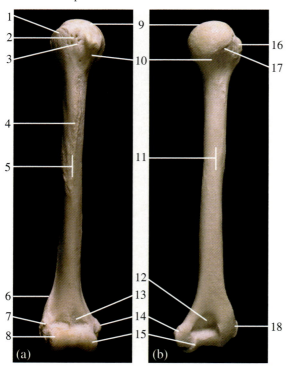

Figure 8.21 The right humerus. (a) An anterior view and (b) a posterior view.

1. Greater tubercle
2. Intertubercular groove
3. Lesser tubercle
4. Deltoid tuberosity
5. Anterior surface of humerus
6. Lateral supracondylar ridge
7. Lateral epicondyle
8. Capitulum
9. Head of humerus
10. Surgical neck
11. Posterior surface of humerus
12. Olecranon fossa
13. Coronoid fossa
14. Medial epicondyle
15. Trochlea
16. Greater tubercle
17. Anatomical neck
18. Lateral epicondyle

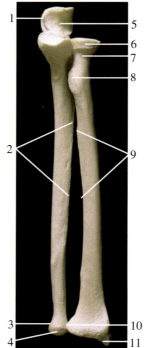

Figure 8.22 An anterior view of the left ulna and radius.

1. Olecranon
2. Interosseous margin
3. Styloid process of ulna
4. Head of ulna
5. Trochlear notch
6. Head of radius
7. Neck of radius
8. Radial tuberosity
9. Interosseous margin
10. Ulnar notch of radius
11. Styloid process of radius

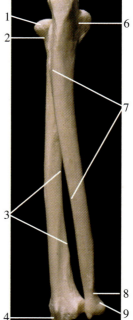

Figure 8.23 A posterior view of the left ulna and radius.

1. Head of radius
2. Neck of radius
3. Interosseous margin
4. Styloid process of radius
5. Olecranon
6. Radial notch of ulna
7. Interosseous margin
8. Head of ulna
9. Styloid process of ulna

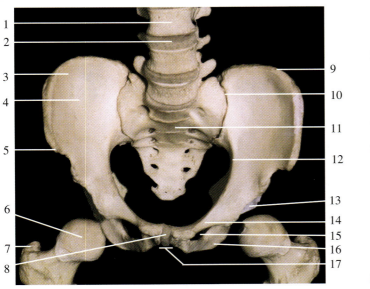

Figure 8.24 An anterior view of the articulated pelvic girdle showing the two coxal bones, the sacrum, and the two femora.

1. Lumbar vertebra
2. Intervertebral disc
3. Ilium
4. Iliac fossa
5. Anterior superior iliac spine
6. Head of femur
7. Greater trochanter
8. Symphysis pubis
9. Crest of the ilium
10. Sacroiliac joint
11. Sacrum
12. Pelvic brim
13. Acetabulum
14. Pubic crest
15. Obturator foramen
16. Ischium
17. Pubic angle

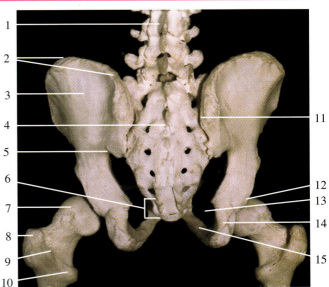

Figure 8.25 A posterior view of the articulated pelvic girdle showing the two coxal bones, the sacrum, and the two femora.

1. Lumbar vertebra
2. Crest of ilium
3. Ilium
4. Sacrum
5. Greater sciatic notch
6. Coccyx
7. Head of femur
8. Greater trochanter
9. Intertrochanteric crest
10. Lesser trochanter
11. Sacroiliac joint
12. Acetabulum
13. Obturator foramen
14. Ischium
15. Pubis

Figure 8.26 The left femur. (a) An anterior view and (b) a posterior view.

1. Fovea capitis femoris
2. Head
3. Neck
4. Lesser trochanter
5. Medial epicondyle
6. Patellar surface
7. Greater trochanter
8. Intertrochanteric crest
9. Intertrochanteric line
10. Lateral epicondyle
11. Lateral condyle
12. Intercondylar fossa
13. Head
14. Fovea capitis femoris
15. Neck
16. Lesser trochanter
17. Linea aspera on shaft (body) of femur
18. Medial epicondyle
19. Medial condyle

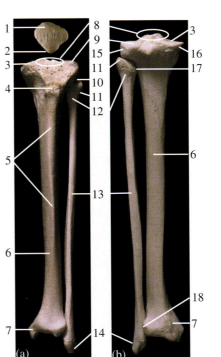

Figure 8.27 An anterior view of the (a) left patella, tibia, and fibula. (b) A posterior view of the left tibia and fibula.

1. Base of patella
2. Apex of patella
3. Medial condyle
4. Tibial tuberosity
5. Anterior crest of tibia
6. Body (shaft) of tibia
7. Medial malleolus
8. Intercondylar tubercles
9. Lateral condyle
10. Tibial articular facet of fibula
11. Head of fibula
12. Neck of fibula
13. Body (shaft) of fibula
14. Lateral malleolus
15. Lateral epicondyle of tibia
16. Medial epicondyle of tibia
17. Fibular articular facet of tibia
18. Fibular notch of tibia

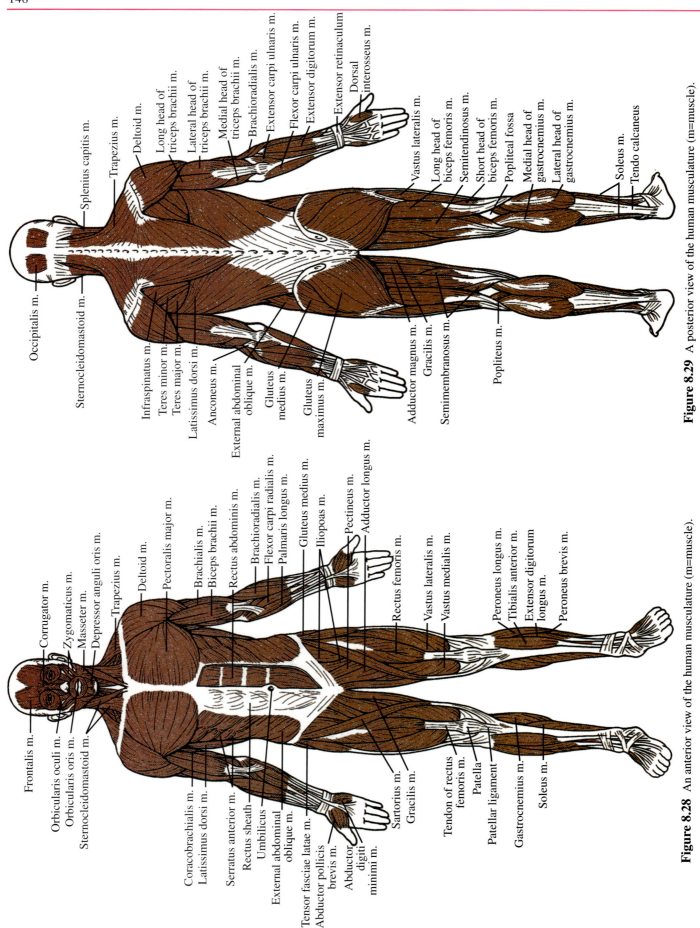

Occipitalis m.

Sternocleidomastoid m.

Splenius capitis m.

Trapezius m.

Deltoid m.

Long head of triceps brachii m.

Lateral head of triceps brachii m.

Medial head of triceps brachii m.

Brachioradialis m.

Extensor carpi ulnaris m.

Flexor carpi ulnaris m.

Extensor digitorum m.

Extensor retinaculum

Dorsal interosseus m.

Vastus lateralis m.

Long head of biceps femoris m.

Semitendinosus m.

Short head of biceps femoris m.

Popliteal fossa

Medial head of gastrocnemius m.

Lateral head of gastrocnemius m.

Soleus m.

Tendo calcaneus

Infraspinatus m.

Teres minor m.

Teres major m.

Latissimus dorsi m.

Anconeus m.

External abdominal oblique m.

Gluteus medius m.

Gluteus maximus m.

Adductor magnus m.

Gracilis m.

Semimembranosus m.

Popliteus m.

Figure 8.29 A posterior view of the human musculature (m=muscle).

Frontalis m.

Corrugator m.

Orbicularis oculi m.

Zygomaticus m.

Orbicularis oris m.

Masseter m.

Depressor anguli oris m.

Sternocleidomastoid m.

Trapezius m.

Deltoid m.

Pectoralis major m.

Brachialis m.

Biceps brachii m.

Rectus abdominis m.

Brachioradialis m.

Flexor carpi radialis m.

Palmaris longus m.

Gluteus medius m.

Iliopoas m.

Pectineus m.

Adductor longus m.

Rectus femoris m.

Vastus lateralis m.

Vastus medialis m.

Peroneus longus m.

Tibialis anterior m.

Extensor digitorum longus m.

Peroneus brevis m.

Coracobrachialis m.

Latissimus dorsi m.

Serratus anterior m.

Rectus sheath

Umbilicus

External abdominal oblique m.

Tensor fasciae latae m.

Abductor pollicis brevis m.

Abductor digiti minimi m.

Sartorius m.

Gracilis m.

Tendon of rectus femoris m.

Patella

Patellar ligament

Gastrocnemius m.

Soleus m.

Figure 8.28 An anterior view of the human musculature (m=muscle).

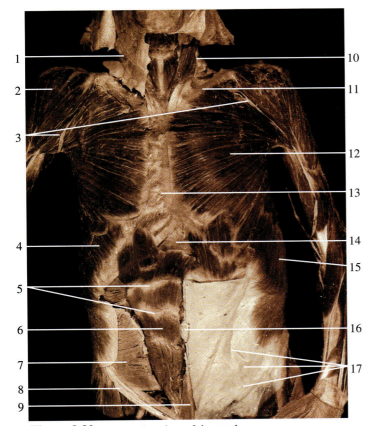

Figure 8.30 An anterior view of the trunk.

1. Platysma m.
2. Deltoid m.
3. Cephalic veins
4. External abdominal oblique m. (aponeurosis removed)
5. Tendinous inscriptions of rectus abdominis m.
6. Rectus abdominis m.
7. Internal abdominal oblique m.
8. Inguinal ligament
9. Pyramidalis m.
10. Sternocleidomastoid m.
11. Clavicle
12. Pectoralis major m.
13. Sternum
14. Xiphoid process
15. External abdominal oblique m.
16. Umbilicus
17. Aponeurosis of external abdominal oblique m.

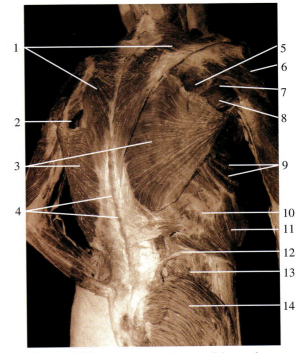

Figure 8.31 A posterolateral view of the trunk.

1. Trapezius m.
2. Triangle of ausculation
3. Latissimus dorsi m.
4. Vertebral column (spinous processes)
5. Infraspinatus m.
6. Deltoid m.
7. Teres minor m.
8. Teres major m.
9. Serratus anterior m.
10. Rib
11. External abdominal oblique m.
12. Iliac crest
13. Gluteus medius m.
14. Gluteus maximus m.

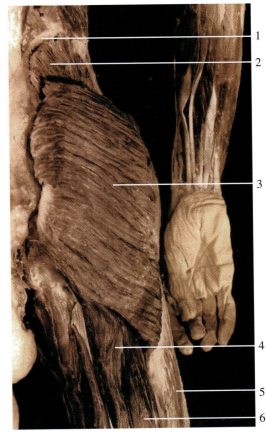

Figure 8.32 The superficial muscles of the gluteal region.

1. Iliac crest
2. Gluteus medius m.
3. Gluteus maximus m.
4. Semitendinosus m.
5. Vastus lateralis m.
6. Biceps femoris m. (long head)

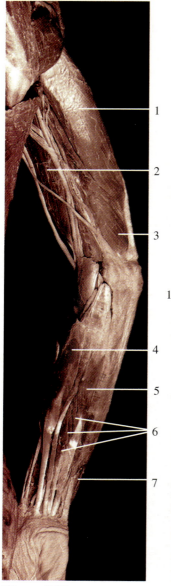

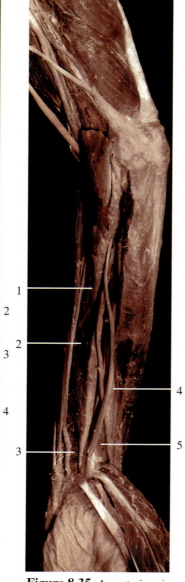

Figure 8.33 The medial brachium and superficial flexors of the right antebrachium.

1. Triceps brachii m. (long head)
2. Biceps brachii m. (short head)
3. Triceps brachii m. (medial head)
4. Flexor carpi radialis m.
5. Palmaris longus m.
6. Superficial digital flexor m.
7. Flexor carpi ulnaris m.

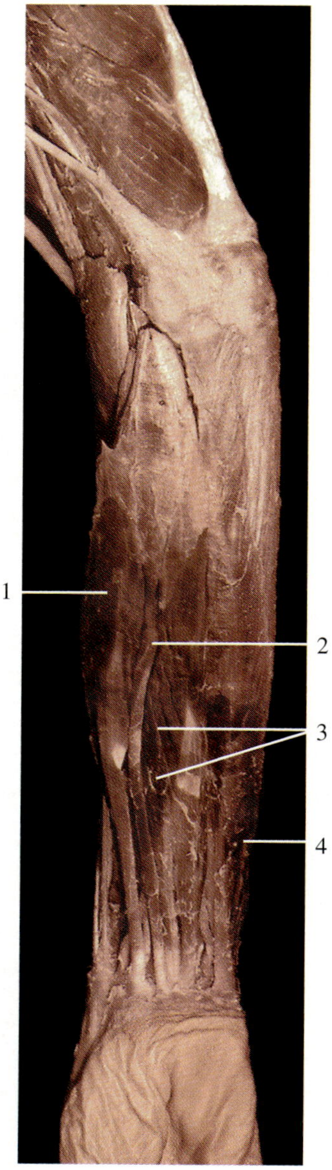

Figure 8.34 An anterior view of the superficial muscles of the right forearm.

1. Flexor carpi radialis m.
2. Palmaris longus m.
3. Superficial digital flexor m.
4. Flexor carpi ulnaris m.

Figure 8.35 An anterior view of the deep muscles of the right forearm.

1. Pronator teres m.
2. Flexor pollicus longus m.
3. Pronator quadratus m.
4. Median nerve
5. Deep digital flexor m.

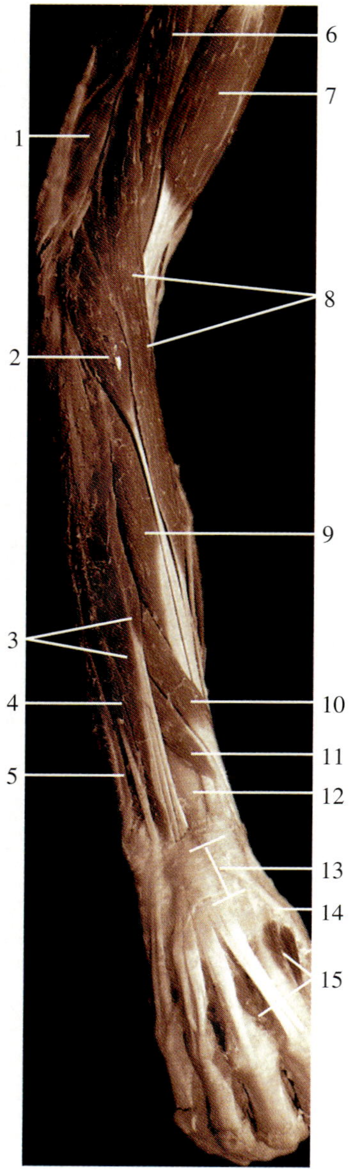

Figure 8.36 A posterior view of the superficial muscles of the right forearm.

1. Triceps brachii m. (medial head)
2. Extensor carpi radialis longus m.
3. Extensor digitorum m.
4. Extensor carpi minimi m.
5. Extensor carpi carpi ulnaris m.
6. Brachialis m.
7. Biceps brachii m. (long head)
8. Brachioradialis m.
9. Extensor carpi radialis brevis m.
10. Abductor pollicis longus m.
11. Extensor pollicus brevis m.
12. Radius
13. Extensor retinaculum
14. Tendon of extensor pollicus longus m.
15. Dorsal interosseous mm.

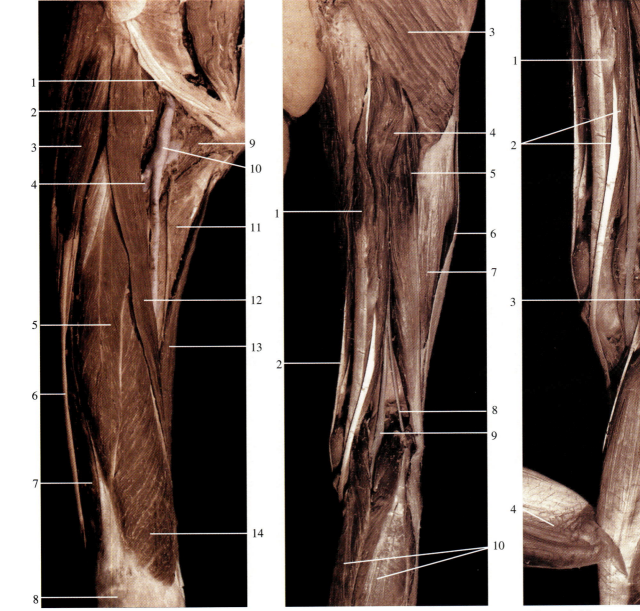

Figure 8.37 An anterior view of the right thigh.

1. Inguinal ligament
2. Iliopsoas m.
3. Tensor faciae latae m.
4. Deep femoral artery
5. Rectus femoris m.
6. Iliotibial tract
7. Vastus lateralis m.
8. Patella
9. Pectineus m.
10. Femoral artery
11. Adductor longus m.
12. Sartorius m.
13. Gracilis m.
14. Vastus medialis m.

Figure 8.38 A posterior view of the right thigh.

1. Semimembranosus m.
2. Gracilis m.
3. Gluteus maximus m.
4. Semitendinosus m.
5. Biceps femoris (long head)
6. Iliotibial tract
7. Vastus lateralis m.
8. Common peroneal nerve
9. Tibial nerve
10. Gastrocnemius m. (lateral and medial heads)

Figure 8.39 The right popliteal fossa and the surrounding structures.

1. Semimembranosus m.
2. Semitendinosus m
3. Plantaris m.
4. Gastrocnemius m. (reflected)
5. Biceps femoris m.
6. Common peroneal nerve
7. Popliteus m.
8. Soleus m.

Controlling systems and sensory organs

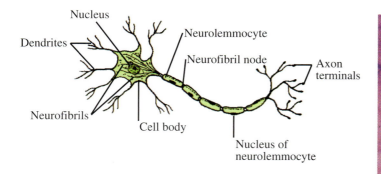

Figure 8.40 Structure of a myelinated neuron.

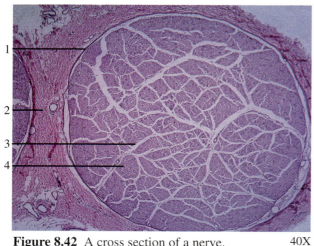

Figure 8.42 A cross section of a nerve.   40X

1. Perineurium
2. Epineurium
3. Endoneurium
4. Bundle of axons

Figure 8.41 The histology of a neuron.   400X

1. Nuclei of surrounding
   neuroglial cells
2. Nucleus of neuron
3. Nucleolus of neuron
4. Dendrites of neuron

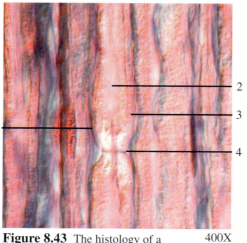

Figure 8.43 The histology of a   400X
myelinated nerve.

1. Neurolemmal sheath
2. Axon
3. Myelin layer
4. Neurofibril node
   (node of Ranvier)

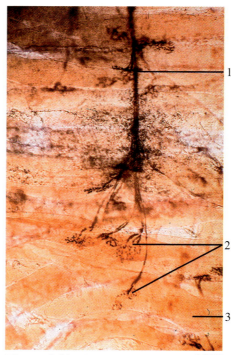

Figure 8.44 The neuromuscular   250X
junction.

1. Motor nerve
2. Motor end plates
3. Skeletal muscle fiber

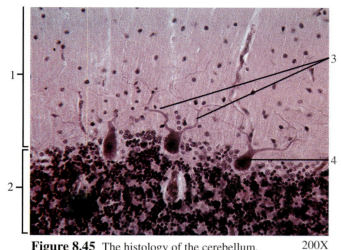

Figure 8.45 The histology of the cerebellum.   200X

1. Molecular layer of cerebellar cortex
2. Granular layer of cerebellar cortex
3. Dendrites of Purkinje cell
4. Purkinje cell body

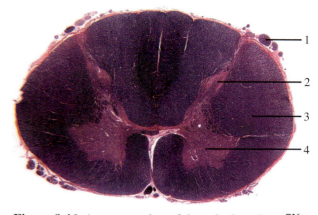

Figure 8.46 A cross section of the spinal cord.   7X

1. Posterior (dorsal) root of spinal nerve
2. Posterior (dorsal) horn (gray matter)
3. Spinal cord tract (white matter)
4. Anterior (ventral) horn (gray matter)

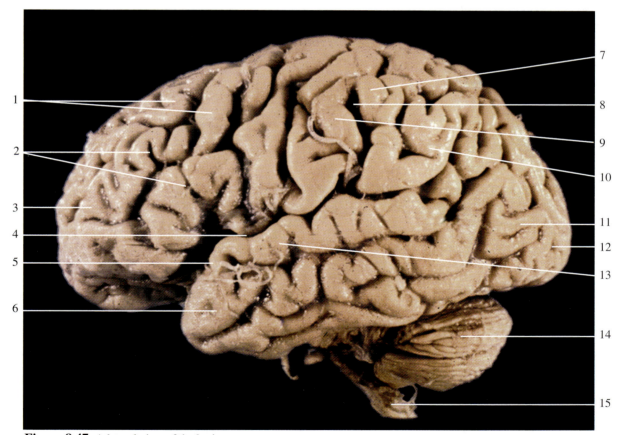

Figure 8.47 A lateral view of the brain.

1. Gyri
2. Sulci
3. Frontal lobe of cerebrum
4. Lateral sulcus
5. Olfactory cerebral cortex
6. Temporal lobe of cerebrum
7. Primary sensory cerebral cortex
8. Central sulcus
9. Primary motor cerebral cortex
10. Parietal lobe of cerebrum
11. Occipital lobe of cerebrum
12. Visual cerebral cortex
13. Auditory cerebral cortex
14. Cerebellum
15. Medulla oblongata

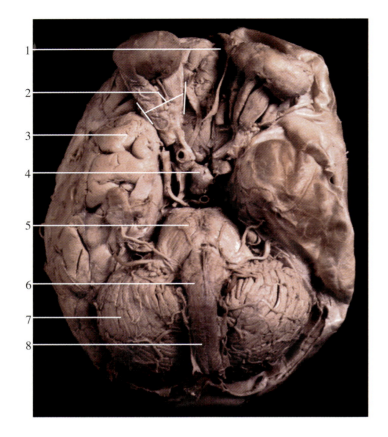

Figure 8.48 An inferior view of the brain with the eyes and part of the meninges still intact.

1. Longitudinal cerebral fissure
2. Muscles of the eye
3. Temporal lobe of cerebrum
4. Pituitary gland
5. Pons
6. Medulla oblongata
7. Cerebellum
8. Spinal cord

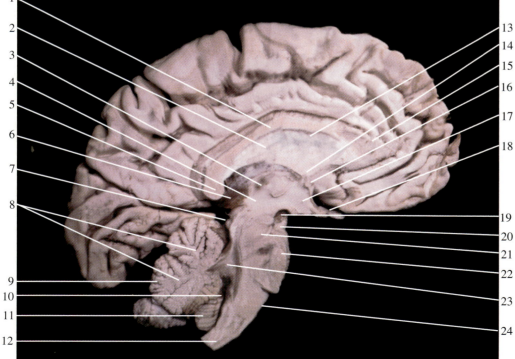

Figure 8.49 A sagittal view of the brain.

1. Truncus of corpus callosum
2. Crus of fornix
3. Third ventricle
4. Posterior commissure
5. Splenium of corpus callosum
6. Pineal body
7. Inferior colliculus
8. Arbor vitae of cerebellum
9. Vermis of cerebellum
10. Choroid plexus of fourth ventricle
11. Tonsilla of cerebellum
12. Medulla oblongata
13. Septum pellucidum (cut)
14. Intraventricular foramen
15. Genu of corpus callosum
16. Anterior commissure
17. Hypothalmus
18. Optic chiasma
19. Oculomotor nerve
20. Cerebral peduncle
21. Midbrain
22. Pons
23. Fourth ventricle
24. Pyramid

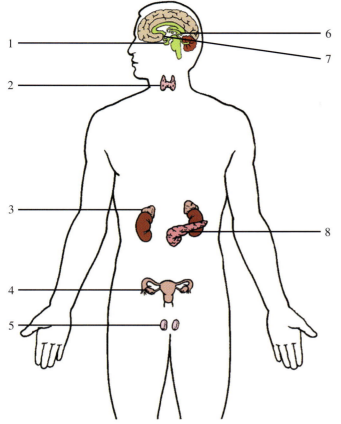

Figure 8.50 Principal endocrine glands.

1. Pituitary gland   5. Testis
2. Thyroid gland   6. Pineal body
3. Adrenal gland   7. Hypothalamus
4. Ovary   8. Pancreas

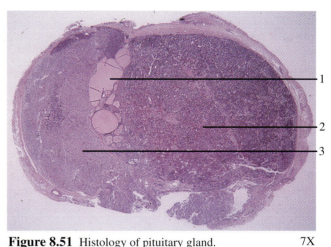

Figure 8.51 Histology of pituitary gland.   7X

1. Pars intermedia (adenohypophysis)
2. Pars distalis (adenohypophysis)
3. Pars nervosa (neurohypophysis)

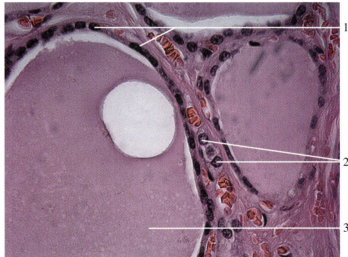

Figure 8.52 Thyroid gland.

1. Follicle cells   400X
2. C cells
3. Colloid within follicle

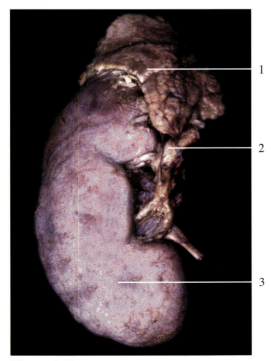

Figure 8.53 The adrenal gland.

1. Adrenal gland   3. Kidney
2. Inferior suprarenal artery

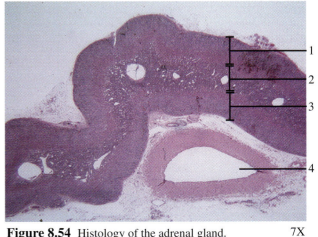

Figure 8.54 Histology of the adrenal gland.   7X

1. Adrenal cortex   3. Adrenal cortex
2. Adrenal medulla   4. Blood vessel

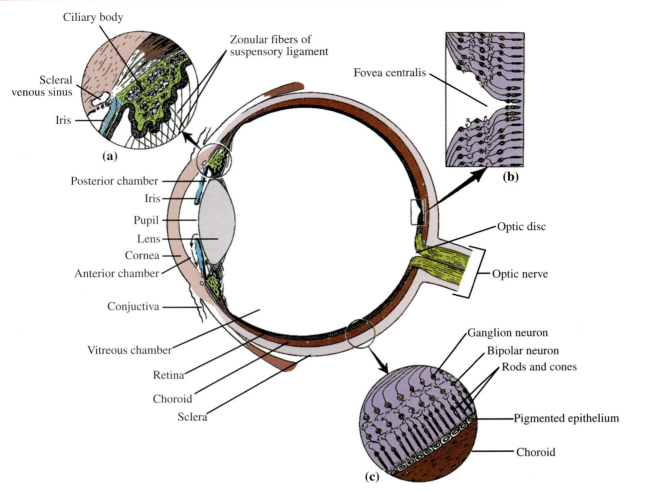

Figure 8.55 Structure of the eye. (a) The ciliary body, (b) fovea centralis, and (c) retina.

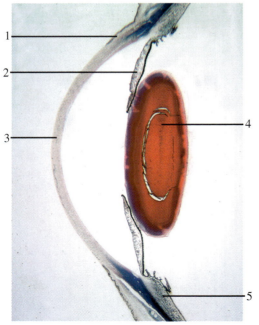

Figure 8.56 Anterior portion of the eye.   7X

1. Conjunctivia   4. Lens
2. Iris   5. Ciliary body
3. Cornea

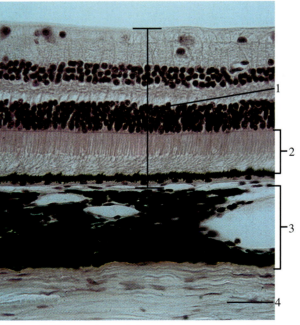

Figure 8.57 Histology of the eye layers.   250X

1. Retina   3. Choroid
2. Rods and cones   4. Sclera

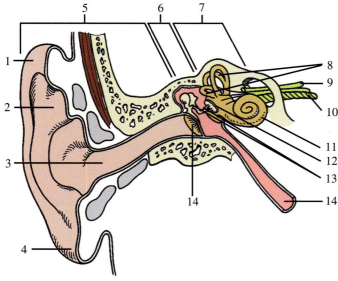

Figure 8.58 Structure of the ear.

1. Helix
2. Auricle
3. External auditory canal
4. Earlobe
5. Outer ear
6. Middle ear
7. Inner ear
8. Semicircular canals
9. Facial nerve
10. Vestibulocochlear nerve
11. Cochlea
12. Vestibular (oval) window
13. Auditory ossicles
14. Auditory tube
15. Tympanic membrane

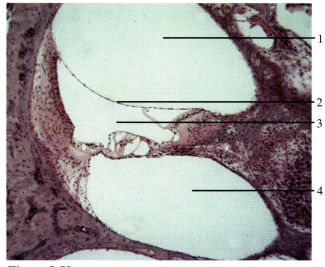

Figure 8.59 Histology of the cochlea.   75X

1. Scala vestibuli
2. Vestibular membrane
3. Cochlear duct
4. Scala tympani

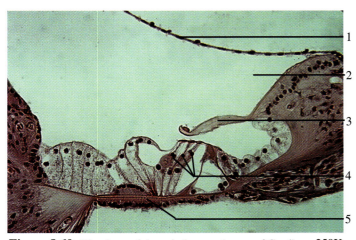

Figure 8.60 Histology of the spiral organ (organ of Corti).   250X

1. Vestibular membrane
2. Cochlear duct
3. Tectorial membrane
4. Hair cells
5. Basilar membrane

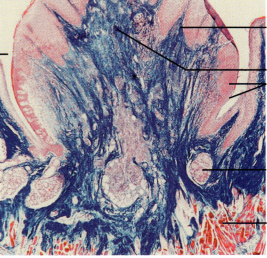
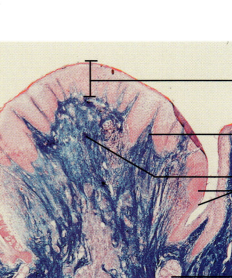

Figure 8.61 Histology of a vallate papilla.   32X

1. Circular furrow
2. Stratified squamous epithelium
3. Secondary papilla
4. Lamina propria
5. Taste buds
6. Serous acini of glands (von Ebner's)
7. Skeletal muscle fibers

Cardiovascular System

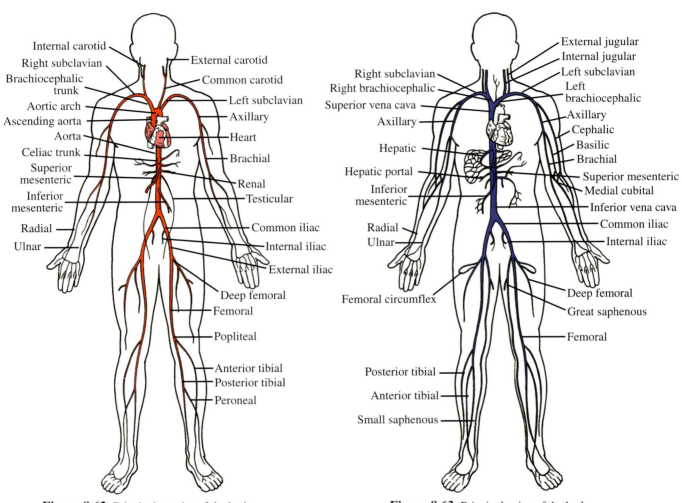

Internal carotid
Right subclavian
Brachiocephalic trunk
Aortic arch
Ascending aorta
Aorta
Celiac trunk
Superior mesenteric
Inferior mesenteric
Radial
Ulnar

External carotid
Common carotid
Left subclavian
Axillary
Heart
Brachial
Renal
Testicular
Common iliac
Internal iliac
External iliac
Deep femoral
Femoral
Popliteal
Anterior tibial
Posterior tibial
Peroneal

Figure 8.62 Principal arteries of the body.

Right subclavian
Right brachiocephalic
Superior vena cava
Axillary
Hepatic
Hepatic portal
Inferior mesenteric
Radial
Ulnar

External jugular
Internal jugular
Left subclavian
Left brachiocephalic
Axillary
Cephalic
Basilic
Brachial
Superior mesenteric
Medial cubital
Inferior vena cava
Common iliac
Internal iliac

Femoral circumflex

Deep femoral
Great saphenous
Femoral

Posterior tibial
Anterior tibial
Small saphenous

Figure 8.63 Principal veins of the body.

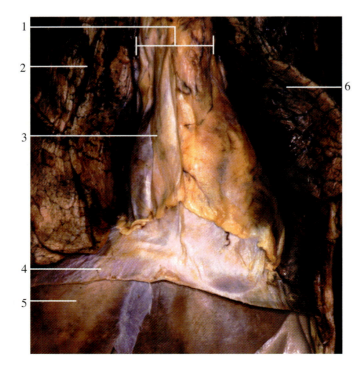

Figure 8.64 The position of the heart within the pericardium.

1. Mediastinum
2. Right lung
3. Pericardium
4. Diaphragm

5. Liver
6. Left lung

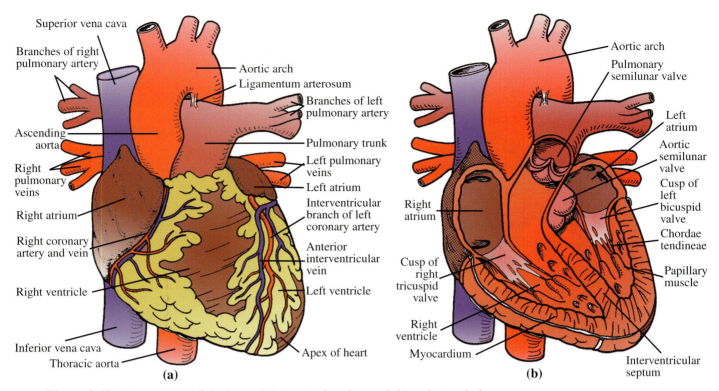

Figure 8.65 The structure of the heart. (a) An anterior view and (b) an internal view.

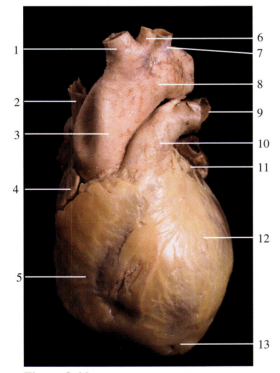

Figure 8.66 An anterior view of the heart and great vessels.

1. Brachiocephalic trunk
2. Superior vena cava
3. Ascending aorta
4. Right atrium
5. Right ventricle
6. Left common carotid artery
7. Left subclavian artery
8. Aortic arch
9. Pulmonary artery
10. Pulmonary trunk
11. Left atrium
12. Left ventricle
13. Apex of heart

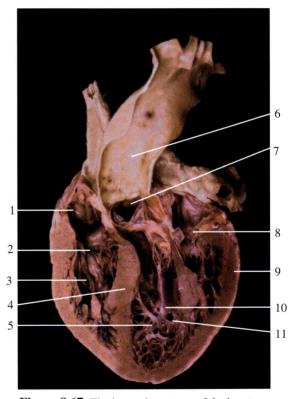

Figure 8.67 The internal structure of the heart.

1. Right atrium
2. Tricuspid valve
3. Right ventricle
4. Interventricular septum
5. Trabeculae carneae
6. Ascending aorta
7. Aortic semilunar valve
8. Bicuspid valve
9. Myocardium
10. Papillary muscle
11. Left ventricle

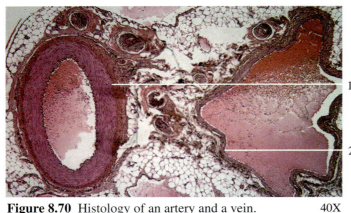

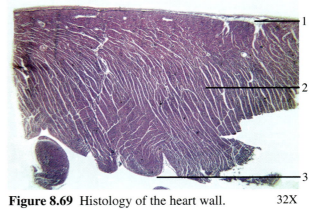

Figure 8.69 Histology of the heart wall.   32X

1. Epicardium   3. Endocardium
2. Myocardium

Figure 8.68 Normal electrocardiogram (ECG).

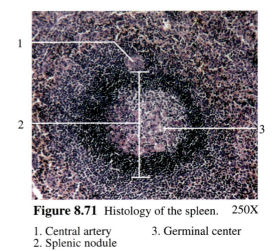

Figure 8.71 Histology of the spleen.   250X

1. Central artery   3. Germinal center
2. Splenic nodule

Figure 8.70 Histology of an artery and a vein.   40X

1. Artery
2. Vein

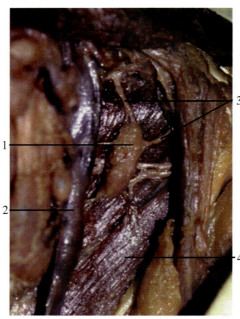

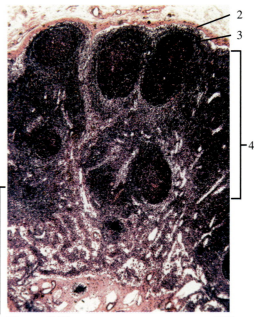

Figure 8.72 A lymph node.

   1. Lymph node   3. Lymphatic vessels
   2. Vein   4. Muscle

Figure 8.73 Histology of a lymph node. 40X

   1. Medulla of lymph node   3. Lymphatic nodule
   2. Capsule   4. Cortex of lymph node

Respiratory system

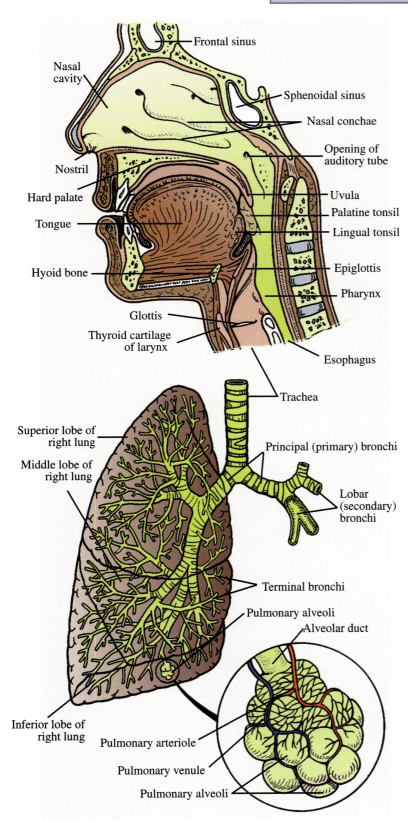

Figure 8.74 The structure of the respiratory system.

Labels in Figure 8.74:
- Frontal sinus
- Nasal cavity
- Sphenoidal sinus
- Nasal conchae
- Opening of auditory tube
- Nostril
- Uvula
- Palatine tonsil
- Hard palate
- Lingual tonsil
- Tongue
- Hyoid bone
- Epiglottis
- Pharynx
- Glottis
- Thyroid cartilage of larynx
- Esophagus
- Trachea
- Superior lobe of right lung
- Principal (primary) bronchi
- Middle lobe of right lung
- Lobar (secondary) bronchi
- Terminal bronchi
- Pulmonary alveoli
- Alveolar duct
- Inferior lobe of right lung
- Pulmonary arteriole
- Pulmonary venule
- Pulmonary alveoli

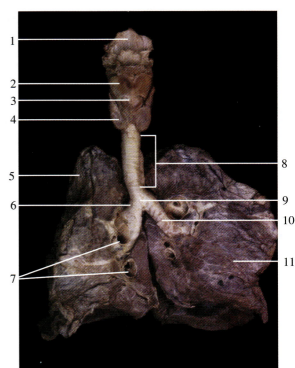

Figure 8.75 An anterior view of the larynx, trachea, and lungs.

1. Epiglottis
2. Thyroid cartilage
3. Cricoid cartilage
4. Thyroid gland
5. Right lung
6. Right principal (primary) bronchus
7. Pulmonary vessels
8. Trachea
9. Carnia
10. Left principal (primary) bronchus
11. Left lung

Figure 8.76 Histology of the trachea. 75X

1. Respiratory epithelium
2. Basement membrane
3. Duct of seromucous gland
4. Seromucous glands
5. Perichondrium
6. Hyaline cartilage

Digestive system

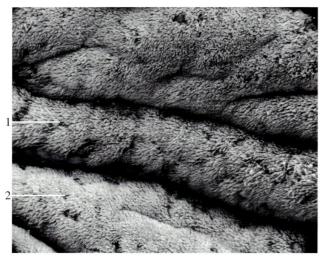

Figure 8.77 An electron micrograph of the lining of the trachea.

1. Cilia
2. Goblet cell

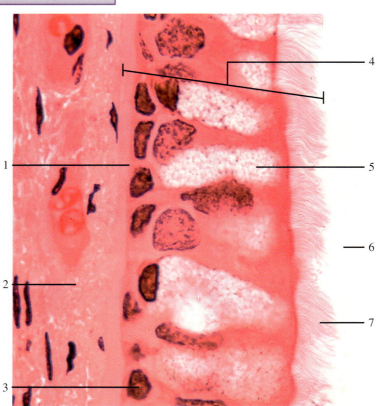

Figure 8.78 Histology of the bronchus.   450X

1. Basement membrane
2. Lamina propria
3. Nucleus
4. Pseudostratified squamous epithelium
5. Goblet cell
6. Lumen of bronchus
7. Cilia

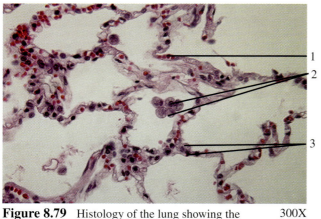

Figure 8.79 Histology of the lung showing the respiratory alveoli.   300X

1. Capillary in alveolar wall
2. Macrophages
3. Type II pneumocytes

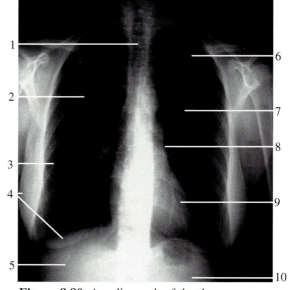

Figure 8.80 A radiograph of the thorax.

1. Thoracic vertebra
2. Right lung
3. Rib
4. Image of right breast
5. Diaphragm/liver
6. Clavicle
7. Left lung
8. Mediastinum
9. Heart
10. Diaphragm/stomach

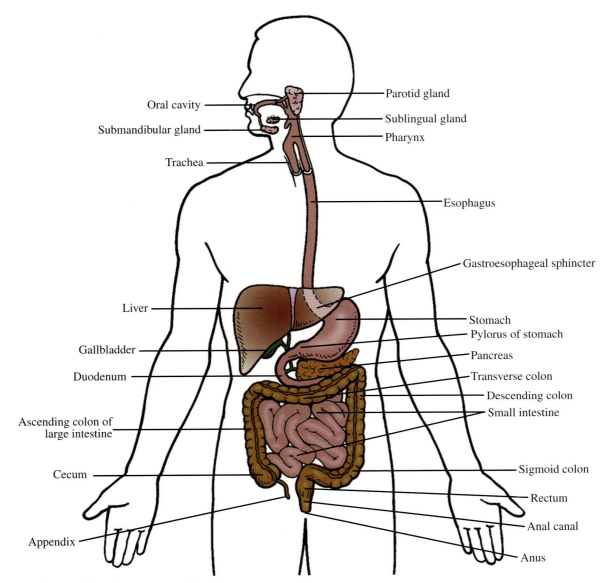

Figure 8.81 The structure of the digestive system.

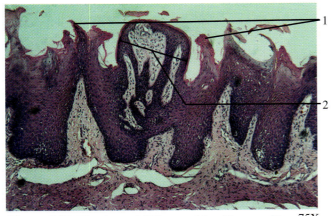

Figure 8.82 Histology of filiform and fungiform papillae. 75X

1. Filiform papillae   2. Fungiform papilla

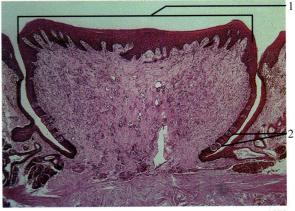

Figure 8.83 Histology of a vallate papilla. 40X

1. Vallate papilla   2. Taste buds

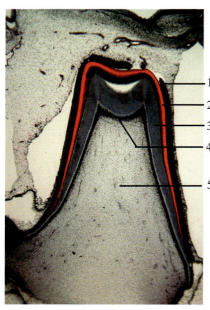

Figure 8.84 Histology of a developing tooth.   40X

1. Ameoblasts
2. Enamel
3. Dentin
4. Odontoblasts
5. Pulp

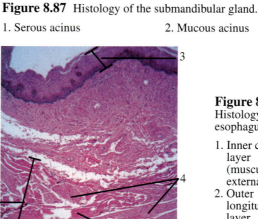

Figure 8.85 Histology of a mature tooth.

1. Dentin (enamel has been dissolved away)
2. Pulp
3. Gingiva
4. Alveolar bone

7X

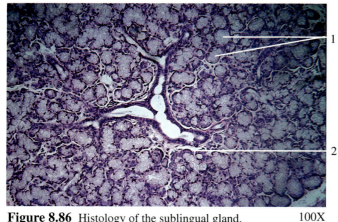

Figure 8.86 Histology of the sublingual gland.   100X

1. Mucous acini   2. Serious demilune

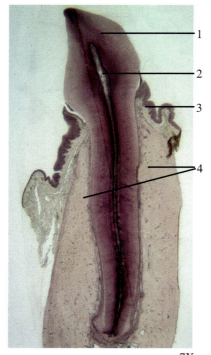

Figure 8.87 Histology of the submandibular gland.   100X

1. Serous acinus   2. Mucous acinus

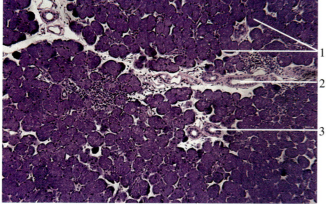

Figure 8.88 Histology of the parotid gland.   250X

1. Serous acini   3. Striated duct
2. Excretory duct

Figure 8.89
Histology of the esophagus.

1. Inner circular layer (muscularis externa)
2. Outer longitudinal layer (muscularis externa)
3. Epithelium
4. Smooth muscle
5. Skeletal muscle

30X

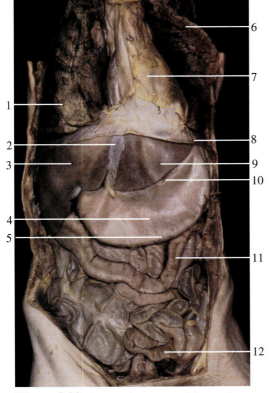

Figure 8.90 An anterior aspect of the trunk.

1. Right lung
2. Falciform ligament
3. Right lobe of liver
4. Body of stomach
5. Greater curvature of stomach
6. Left lung (reflected)
7. Pericardium
8. Diaphragm
9. Left lobe of liver
10. Lesser curvature of stomach
11. Transverse colon
12. Small intestine

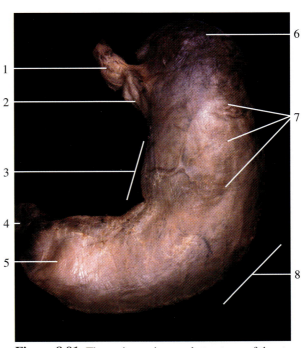

Figure 8.91 The major regions and structures of the stomach.

1. Esophagus
2. Cardiac portion of stomach
3. Lesser curvature of stomach
4. Duodenum
5. Pylorus of stomach
6. Fundus of stomach
7. Body of stomach
8. Greater curvature of stomach

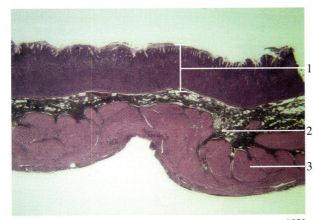

Figure 8.92 Histology of the stomach wall.   10X

1. Mucosa
2. Submucosa
3. Muscularis externa

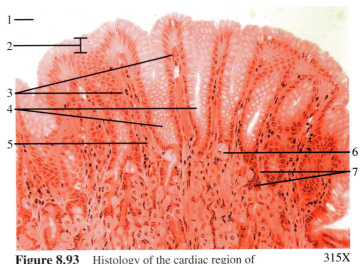

Figure 8.93 Histology of the cardiac region of the stomach.   315X

1. Lumen of stomach
2. Surface epithelium
3. Mucosal ridges
4. Gastric pits
5. Lamina propria
6. Parietal cells
7. Chief (zymogenic) cells

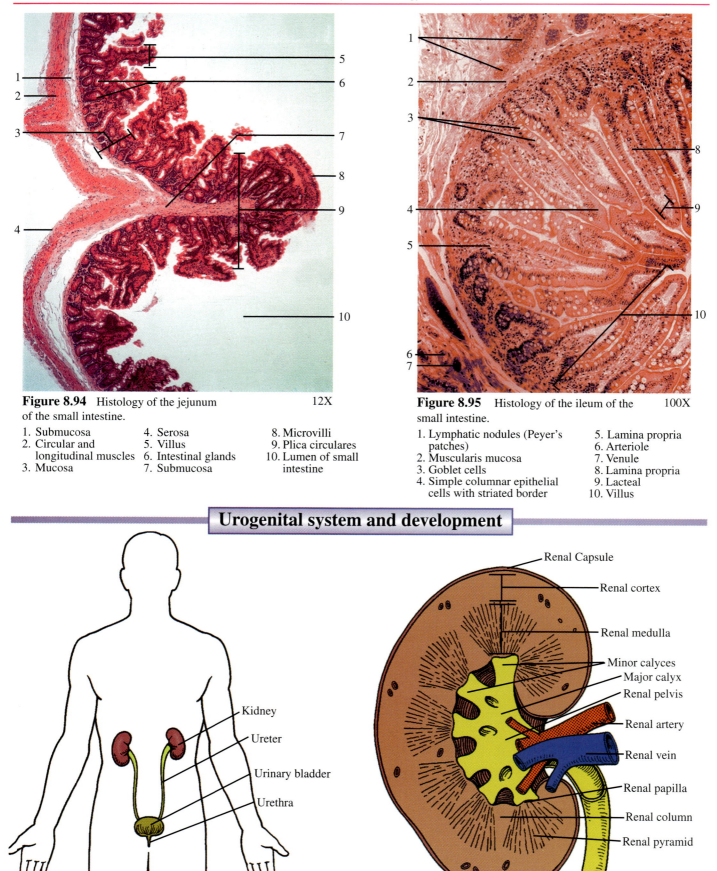

Figure 8.94 Histology of the jejunum   12X
of the small intestine.

1. Submucosa
2. Circular and longitudinal muscles
3. Mucosa
4. Serosa
5. Villus
6. Intestinal glands
7. Submucosa
8. Microvilli
9. Plica circulares
10. Lumen of small intestine

Figure 8.95   Histology of the ileum of the   100X
small intestine.

1. Lymphatic nodules (Peyer's patches)
2. Muscularis mucosa
3. Goblet cells
4. Simple columnar epithelial cells with striated border
5. Lamina propria
6. Arteriole
7. Venule
8. Lamina propria
9. Lacteal
10. Villus

Urogenital system and development

Kidney
Ureter
Urinary bladder
Urethra

Figure 8.96 The organs of the urinary system.

Renal Capsule
Renal cortex
Renal medulla
Minor calyces
Major calyx
Renal pelvis
Renal artery
Renal vein
Renal papilla
Renal column
Renal pyramid
Ureter

Figure 8.97 The structure of the kidney.

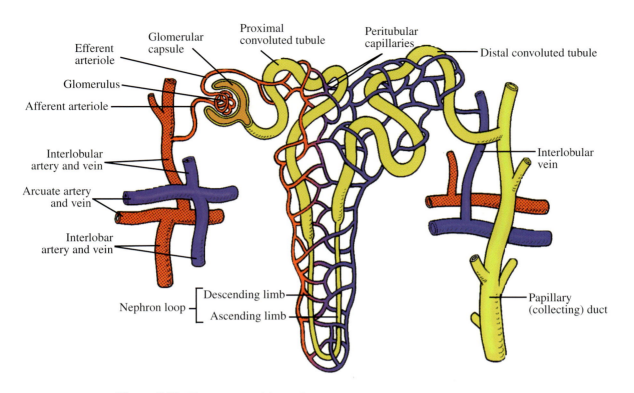

Figure 8.98 The structure of the nephron.

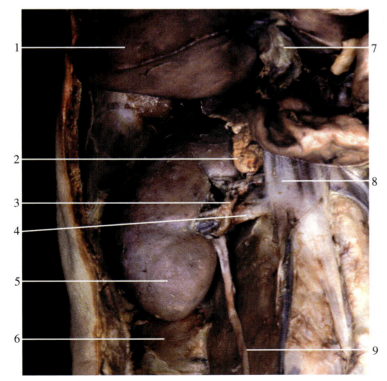

Figure 8.99 The kidney, ureter, and surrounding structures.

1. Liver
2. Adrenal gland
3. Renal artery
4. Renal vein
5. Right kidney
6. Quadratus lumborum muscle
7. Gallbladder
8. Inferior vena cava
9. Ureter

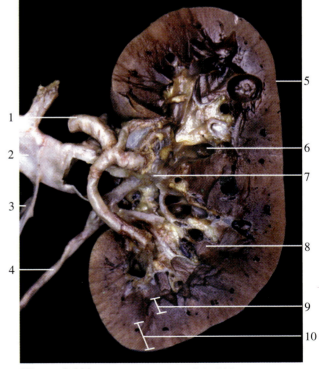

Figure 8.100 A coronal section of the kidney.

1. Renal artery
2. Renal vein
3. Left testicular vein
4. Ureter
5. Capsule
6. Major calyx
7. Renal pelvis
8. Renal papilla
9. Renal medulla
10. Renal cortex

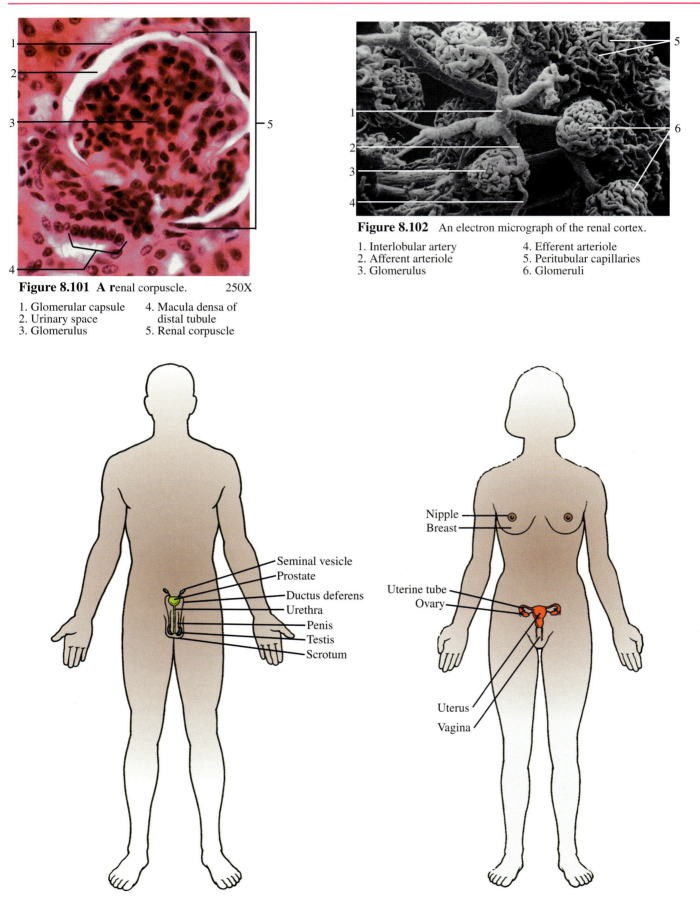

Figure 8.101 A renal corpuscle.	250X

1. Glomerular capsule	4. Macula densa of
2. Urinary space	 distal tubule
3. Glomerulus	5. Renal corpuscle

Figure 8.102 An electron micrograph of the renal cortex.

1. Interlobular artery	4. Efferent arteriole
2. Afferent arteriole	5. Peritubular capillaries
3. Glomerulus	6. Glomeruli

Seminal vesicle
Prostate
Ductus deferens
Urethra
Penis
Testis
Scrotum

Nipple
Breast

Uterine tube
Ovary

Uterus
Vagina

Figure 8.103 The organs of the male reproductive system.

Figure 8.104 The organs of the female reproductive system.

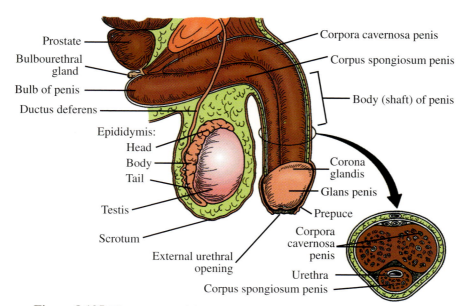

Figure 8.105 The structure of the male genitalia.

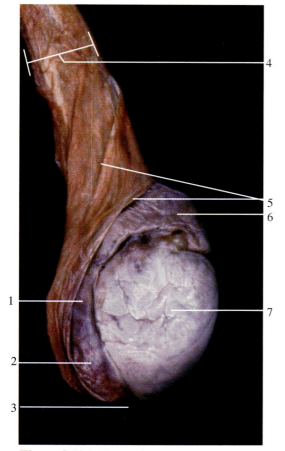

Figure 8.106 The testis and associated structures.

1. Body of epididymis
2. Tail of epididymis
3. Gubernaculum
4. Spermatic cord
5. Spermatic fascia
6. Head of epididymis
7. Testis

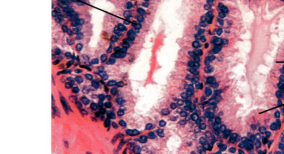

Figure 8.107 Histology of a seminiferous tubule in the testis.   256X

1. Sustentacular (nurse) cells
2. Spermatids
3. Maturing spermatozoa
4. Fibroblasts
5. Primary spermatocytes
6. Secondary spermatocytes
7. Lumen of seminiferous tubule
8. Interstitial cell (of Leydig)
9. Spermatogonium

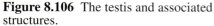

Figure 8.108 Histology of the prostate.   125X

1. Glandular epithelium   2. Smooth muscle   3. Glandular acini

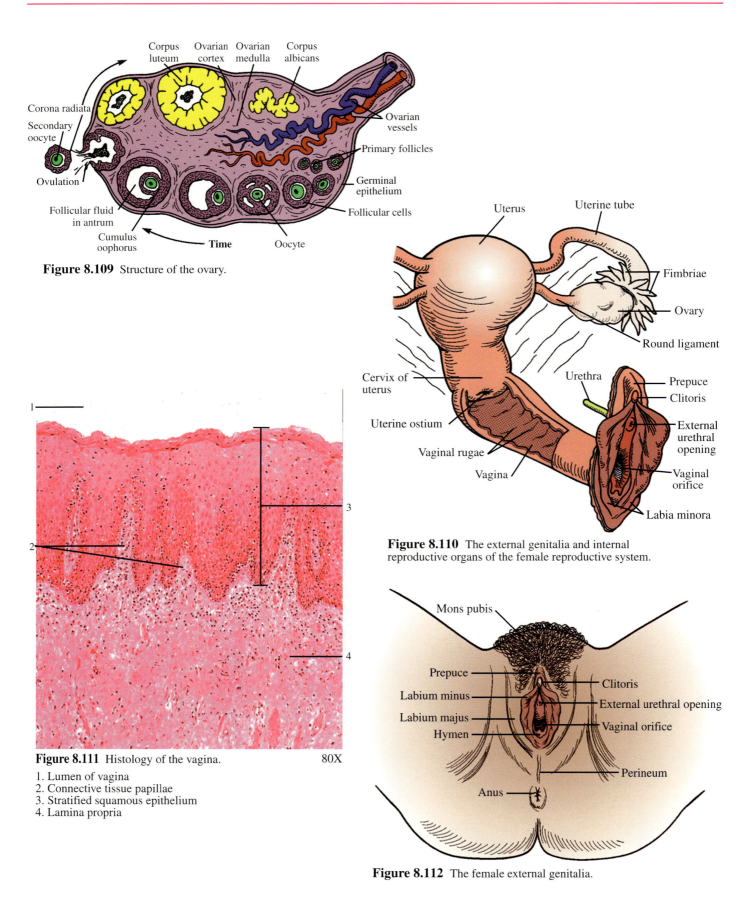

Figure 8.109 Structure of the ovary.

Figure 8.111 Histology of the vagina. 80X

1. Lumen of vagina
2. Connective tissue papillae
3. Stratified squamous epithelium
4. Lamina propria

Figure 8.110 The external genitalia and internal reproductive organs of the female reproductive system.

Figure 8.112 The female external genitalia.

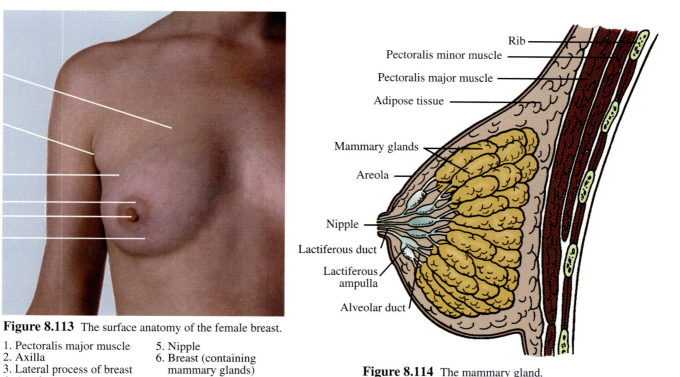

Figure 8.113 The surface anatomy of the female breast.

1. Pectoralis major muscle
2. Axilla
3. Lateral process of breast
4. Areola
5. Nipple
6. Breast (containing mammary glands)

Rib
Pectoralis minor muscle
Pectoralis major muscle
Adipose tissue
Mammary glands
Areola
Nipple
Lactiferous duct
Lactiferous ampulla
Alveolar duct

Figure 8.114 The mammary gland.

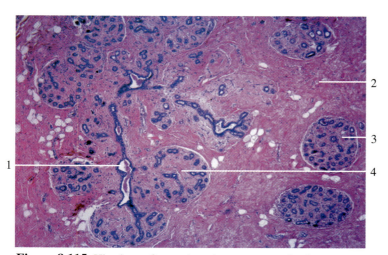

Figure 8.115 Histology of a non-lactating mammary gland.

1. Interlobular duct
2. Interlobular connective tissue
3. Lobule of glandular tissue
4. Intralobular connective tissue

Vertebrate Dissections

9

Class Agnatha

An understanding of the structure of a vertebrate organism is requisite to learning about physiological mechanisms and about how the animal functions in its environment. The selective pressures that determine evolutionary changes frequently have an influence on anatomical structures. Studying dissected specimens, therefore, provides phylogenetic information about how groups of organisms are related.

Some biology laboratories have the resources to provide students with opportunities for doing selected vertebrate dissections. For these students, the photographs contained in this chapter will be a valuable source for identification of structures on your specimens as they are dissected and studied. If dissection specimens are not available, the excellent photographs of carefully dissected prepared specimens presented in this chapter will be an adequate substitute. Care has gone into the preparation of these specimens to depict and identify the principal body structures from representative specimens of each of the classes of vertebrates. Selected human cadaver dissections are shown in photographs contained in chapter 8. As the anatomy of vertebrate specimens is studied in this chapter, observe the photographs of human dissections in the previous chapter and note the similarities of body structure, particularly to those of another mammal.

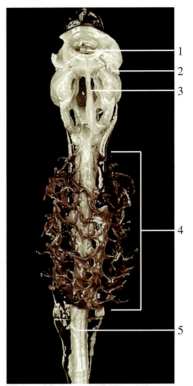

Figure 9.1 A ventral view of the cartilaginous skeleton of the marine lamprey, *Petromyzon marinus*.

1. Buccal cavity
2. Skull
3. Tongue support
4. Branchial basket
5. Notochord

Class Chondrichthyes

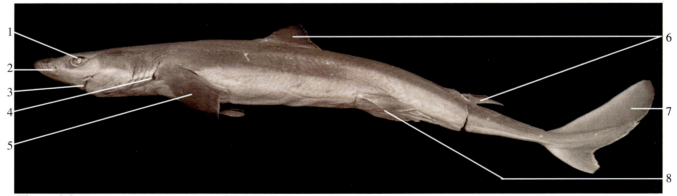

Figure 9.2 The external anatomy of the dogfish shark, *Squalus acanthias*.

1. Eye
2. Nostril
3. Mouth
4. Gill slits
5. Pectoral fin
6. Dorsal fins
7. Caudal fin (heterocercal tail)
8. Pelvic fin

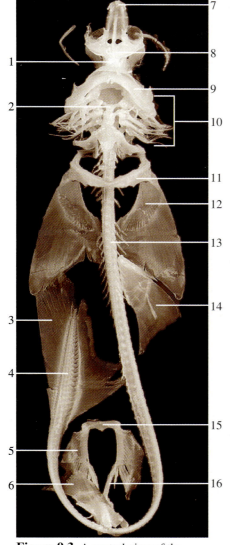

Figure 9.3 A ventral view of the cartilaginous skeleton of a male dogfish shark.

1. Palatopterygoquadrate cartilage (upper jaw)
2. Hypobranchial cartilage
3. Caudal fin
4. Caudal vertebrae
5. Pelvic fin
6. Posterior dorsal fin
7. Rostrum
8. Chondrocranium
9. Meckel's cartilage (lower jaw)
10. Visceral arches
11. Pectoral girdle
12. Pectoral fin
13. Trunk vertebrae
14. Anterior dorsal fin
15. Pelvic girdle
16. Clasper

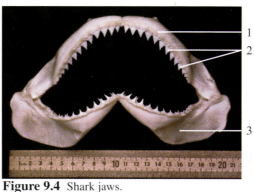

Figure 9.4 Shark jaws.

1. Palatopterygoquadrate cartilage
2. Placoid teeth
3. Meckel's cartilage

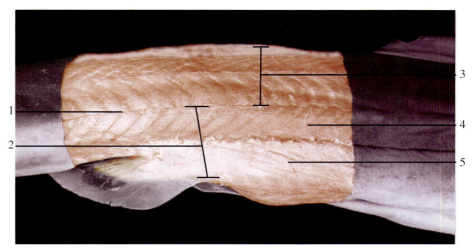

Figure 9.5 A lateral view of the axial musculature of the dogfish shark.

1. Transverse septum
2. Hypaxial myotome portion
3. Epaxial myotome portion
4. Lateral bundle of myotomes
5. Ventral bundle of myotomes

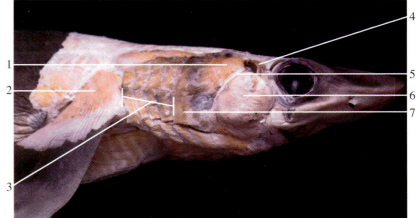

Figure 9.6 The musculature of the jaw, gills, and pectoral fin of a dogfish shark.

1. 2nd dorsal constrictor
2. Levator of pectoral fin
3. 3rd through 6th ventral constrictors
4. Spiracular muscle
5. Facial nerve (hyomandibular branch)
6. Mandibular adductor
7. 2nd ventral constrictor

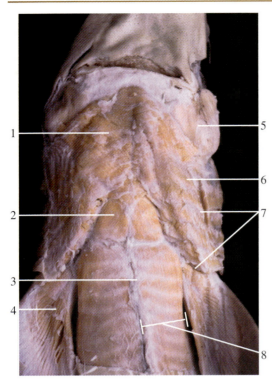

Figure 9.7 A ventral view of the hypobranchial musculature of the dogfish shark.

1. 1st ventral constrictor
2. Common coracoarcual
3. Linea alba
4. Depressor of pectoral fin
5. Mandibular adductor
6. 2nd ventral constrictor
7. 3rd through 6th ventral constrictors
8. Hypaxial muscle

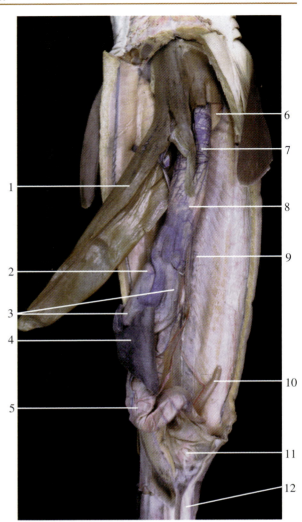

Figure 9.8 The internal anatomy of a male dogfish shark.

1. Right lobe of liver
2. Pyloric sphincter valve
3. Stomach (pyloric region)
4. Spleen
5. Ileum
6. Testis
7. Esophagus
8. Stomach (cardiac region)
9. Kidney
10. Rectal gland
11. Cloaca
12. Clasper

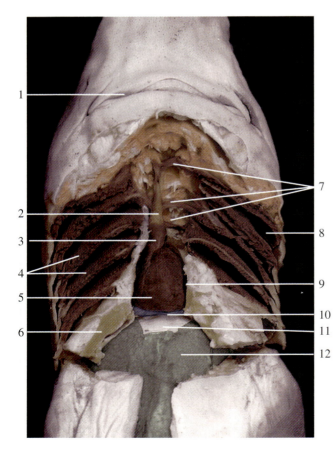

Figure 9.9 The heart, gills, and associated vessels of a dogfish shark.

1. Mouth
2. Ventral aorta
3. Conus ateriosus
4. Gills
5. Ventricle
6. Pectoral girdle (cut)
7. Afferent branchial arteries
8. Gill cleft
9. Pericardial cavity
10. Sinus venosus
11. Transverse septum
12. Liver

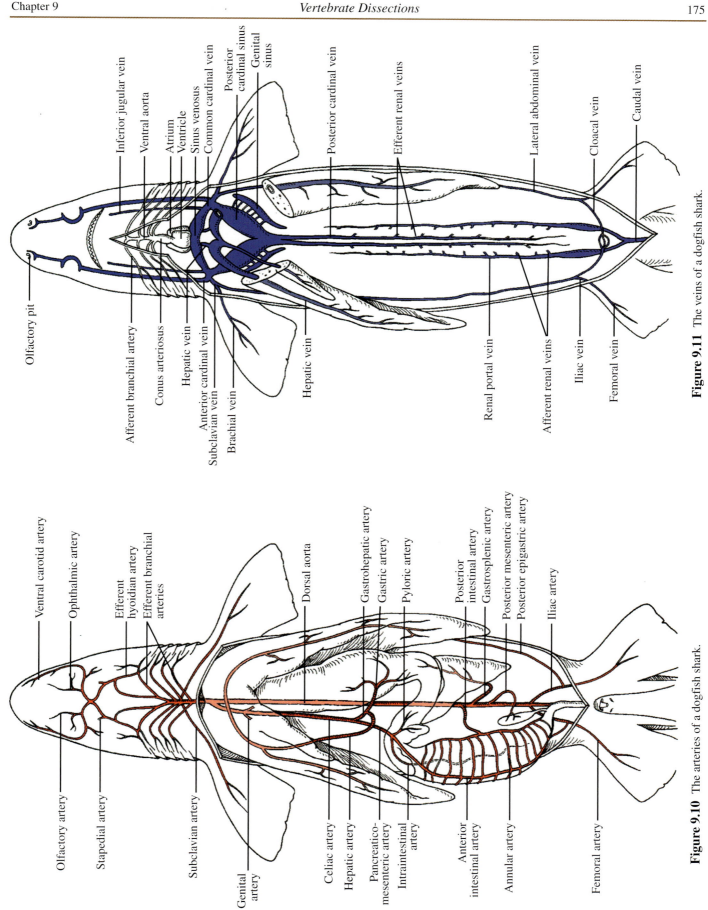

Olfactory pit

Inferior jugular vein
Ventral aorta
Atrium
Ventricle
Sinus venosus
Common cardinal vein
Posterior cardinal sinus
Genital sinus

Posterior cardinal vein

Efferent renal veins

Lateral abdominal vein
Cloacal vein
Caudal vein

Afferent branchial artery
Conus arteriosus
Hepatic vein
Anterior cardinal vein
Subclavian vein
Brachial vein

Hepatic vein

Renal portal vein

Afferent renal veins
Iliac vein
Femoral vein

Figure 9.11 The veins of a dogfish shark.

Ventral carotid artery
Ophthalmic artery

Efferent hyoidian artery
Efferent branchial arteries

Dorsal aorta

Gastrohepatic artery
Gastric artery
Pyloric artery

Posterior intestinal artery
Gastrosplenic artery
Posterior mesenteric artery
Posterior epigastric artery

Iliac artery

Olfactory artery
Stapedial artery

Subclavian artery

Genital artery

Celiac artery
Hepatic artery
Pancreatico-mesenteric artery
Intraintestinal artery

Anterior intestinal artery
Annular artery

Femoral artery

Figure 9.10 The arteries of a dogfish shark.

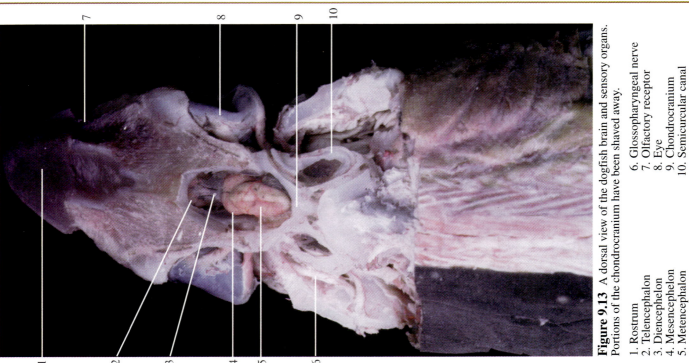

Figure 9.13 A dorsal view of the dogfish brain and sensory organs. Portions of the chondrocranium have been shaved away.

1. Rostrum
2. Telencephalon
3. Diencephelon
4. Mesencephelon
5. Metencephalon
6. Glossopharyngeal nerve
7. Olfactory receptor
8. Eye
9. Chondrocranium
10. Semicurcular canal

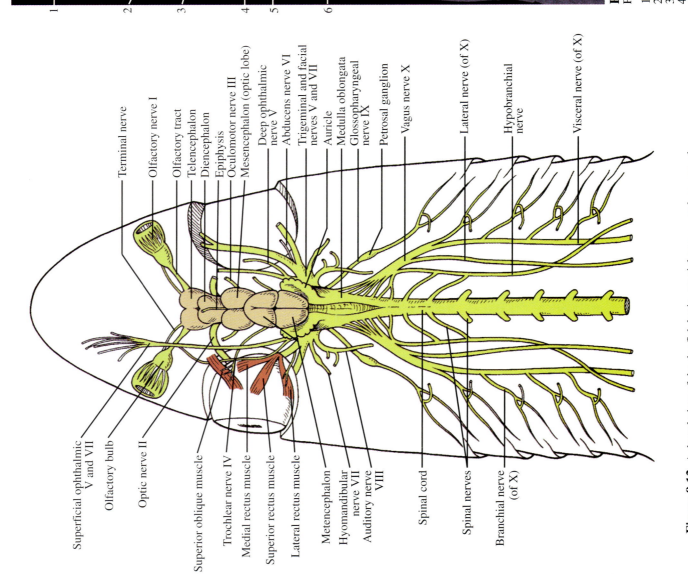

Figure 9.12 A dorsal view of the dogfish brain, cranial nerves, and eye muscles.

Class Osteichthyes

Figure 9.14 The skeleton of a perch.

1. Anterior dorsal fin
2. Posterior dorsal fin
3. Caudal fin
4. Orbit
5. Premaxilla
6. Dentary
7. Branchial skeleton
8. Opercular bones
9. Pectoral girdle
10. Pelvic girdle
11. Pectoral fin
12. Pelvic fin
13. Rib
14. Vertebral column
15. Anal fin
16. Haemal spine
17. Neural spine

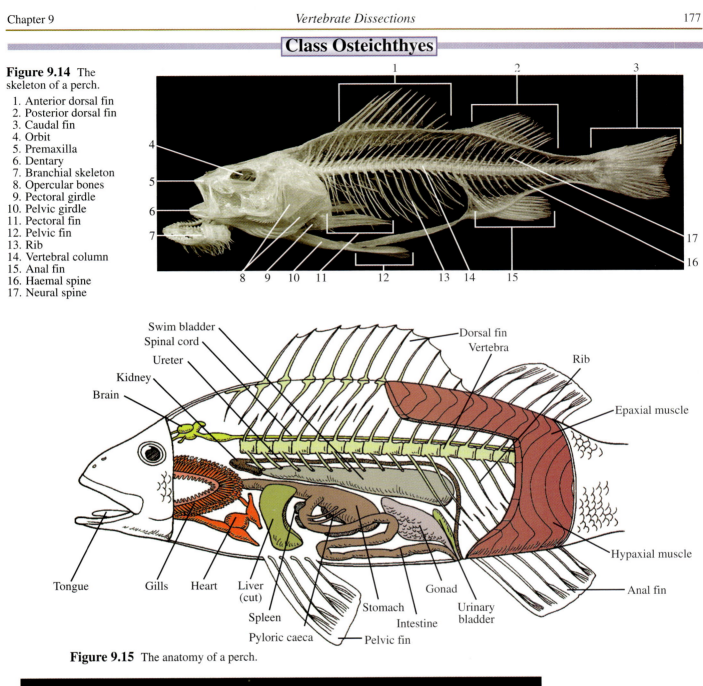

Figure 9.15 The anatomy of a perch.

Figure 9.16

The viscera of a perch.

1. Epaxial muscles
2. Ribs
3. Gill
4. Liver (cut)
5. Heart
6. Pyloric cecum
7. Vertebrae
8. Swim bladder
9. Urinary bladder
10. Gonad
11. Anus
12. Intestine
13. Stomach

Class Amphibia

Figure 9.17 The surface anatomy and body regions of the leopard frog, *Rana pipiens.*

1. Ankle	4. Eyes	7. Brachium
2. Knee	5. Nostril	8. Antebrachium
3. Foot	6. Tympanic membrane	9. Digits

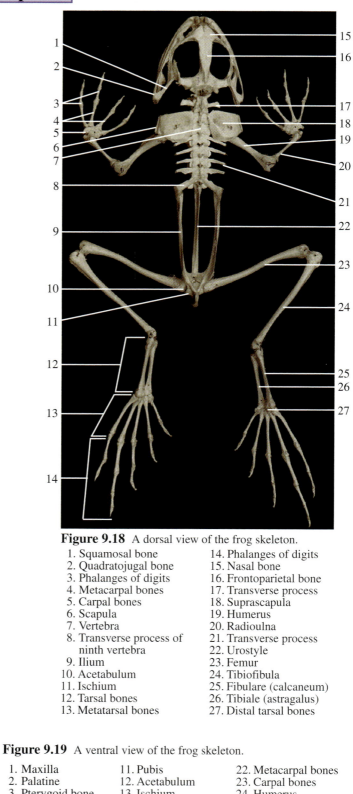

Figure 9.18 A dorsal view of the frog skeleton.

1. Squamosal bone	14. Phalanges of digits
2. Quadratojugal bone	15. Nasal bone
3. Phalanges of digits	16. Frontoparietal bone
4. Metacarpal bones	17. Transverse process
5. Carpal bones	18. Suprascapula
6. Scapula	19. Humerus
7. Vertebra	20. Radioulna
8. Transverse process of	21. Transverse process
ninth vertebra	22. Urostyle
9. Ilium	23. Femur
10. Acetabulum	24. Tibiofibula
11. Ischium	25. Fibulare (calcaneum)
12. Tarsal bones	26. Tibiale (astragalus)
13. Metatarsal bones	27. Distal tarsal bones

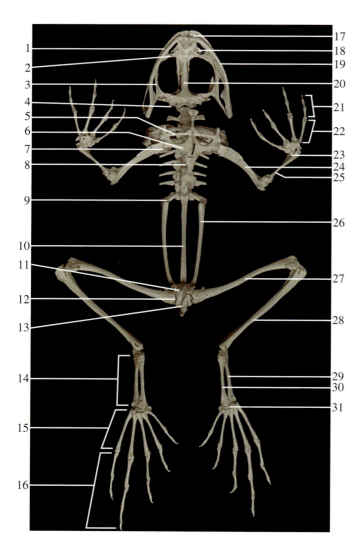

Figure 9.19 A ventral view of the frog skeleton.

1. Maxilla	11. Pubis	22. Metacarpal bones
2. Palatine	12. Acetabulum	23. Carpal bones
3. Pterygoid bone	13. Ischium	24. Humerus
4. Exoccipital bone	14. Tarsal bones	25. Radioulna
5. Clavicle	15. Metatarsal bones	26. Ilium
6. Coracoid	16. Phalanges of digits	27. Femur
7. Glenoid fossa	17. Premaxilla	28. Tibiofibula
8. Sternum	18. Vomer	29. Fibulare
9. Transverse process	19. Dentary	(calcaneum)
of ninth vertebra	20. Parasphenoid bone	30. Tibiale (astragalus)
10. Urostyle	21.Phalanges of digits	31. Distal tarsal bones

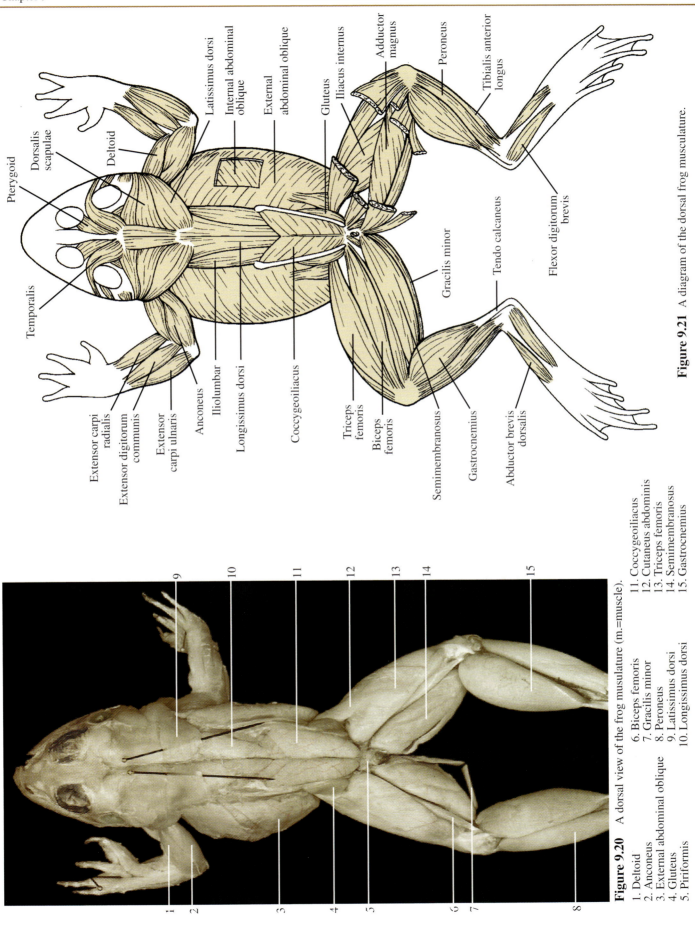

Figure 9.20 A dorsal view of the frog musulature (m.=muscle).

1. Deltoid
2. Anconeus
3. External abdominal oblique
4. Gluteus
5. Piriformis
6. Biceps femoris
7. Gracilis minor
8. Peroneus
9. Latissimus dorsi
10. Longissimus dorsi
11. Coccygeoiliacus
12. Cutaneus abdominis
13. Triceps femoris
14. Semimembranosus
15. Gastrocnemius

Figure 9.21 A diagram of the dorsal frog musculature.

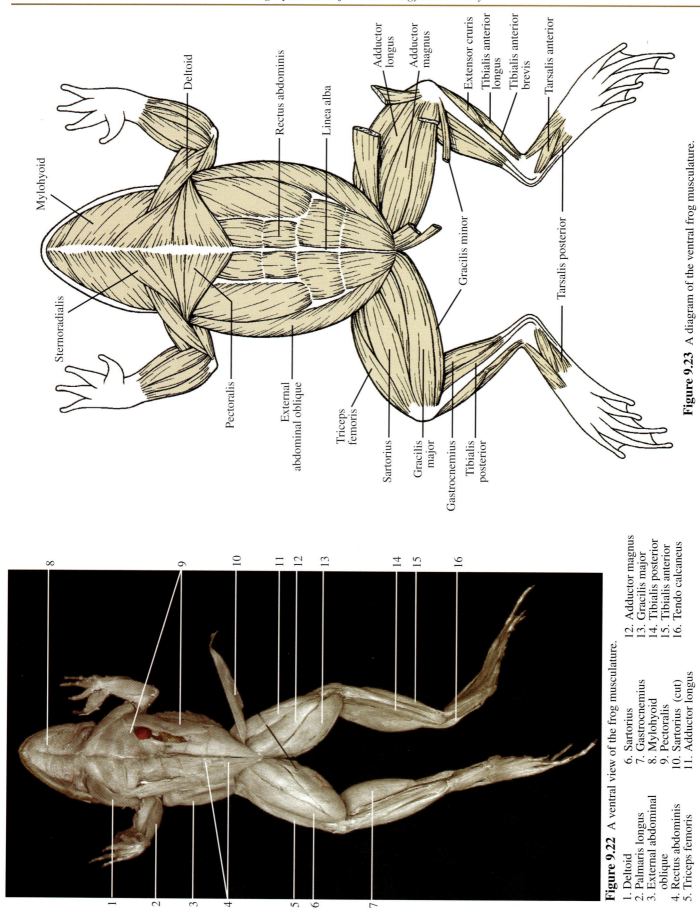

Deltoid

Mylohyoid

Rectus abdominis

Linea alba

Adductor longus

Adductor magnus

Extensor cruris

Tibialis anterior longus

Tibialis anterior brevis

Tarsalis anterior

Sternoradialis

Pectoralis

External abdominal oblique

Gracilis minor

Tarsalis posterior

Triceps femoris

Sartorius

Gracilis major

Gastrocnemius

Tibialis posterior

Figure 9.23 A diagram of the ventral frog musculature.

Figure 9.22 A ventral view of the frog musculature.

1. Deltoid
2. Palmaris longus
3. External abdominal oblique
4. Rectus abdominis
5. Triceps femoris
6. Sartorius
7. Gastrocnemius
8. Mylohyoid
9. Pectoralis
10. Sartorius (cut)
11. Adductor longus
12. Adductor magnus
13. Gracilis major
14. Tibialis posterior
15. Tibialis anterior
16. Tendo calcaneus

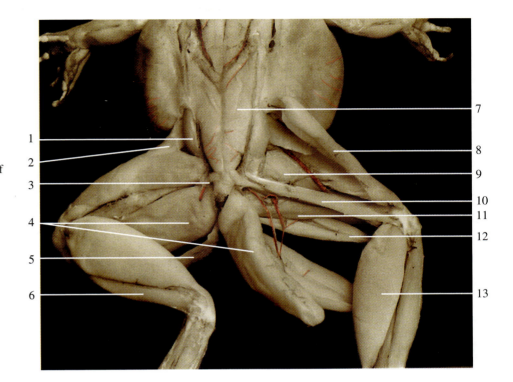

Figure 9.24 A dorsal view of the leg muscles of a frog.

1. Gluteus m.
2. Cutaneus abdominis m.
3. Piriformis m.
4. Semimembranosus m.
5. Gracilis minor m.
6. Peroneus m.
7. Coccygeoiliacus m.
8. Triceps femoris m. (cut)
9. Iliacus internus m.
10. Biceps femoris m.
11. Adductor magnus m.
12. Semitendinosus m .
13. Gastrocnemius m.

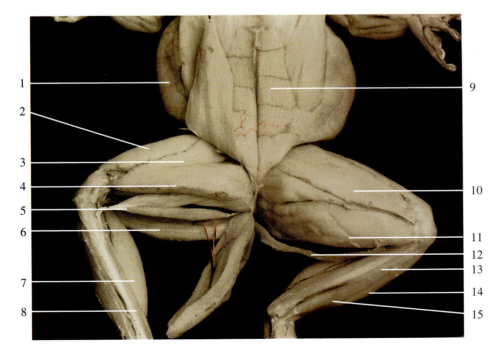

Figure 9.25 A ventral view of the leg muscles of a frog.

1. External abdominal oblique m.
2. Triceps femoris m.
3. Adductor longus m.
4. Adductor magnus m.
5. Semitendinosus m.
6. Semimembranosus m.
7. Gastrocnemius m.
8. Tibialis posterior m.
9. Rectus abdominis m.
10. Sartorius m.
11. Gracilis major m.
12. Gracilis minor m.
13. Extensor cruris m.
14. Tibialis anterior longus m.
15. Tibialis anterior brevis m.

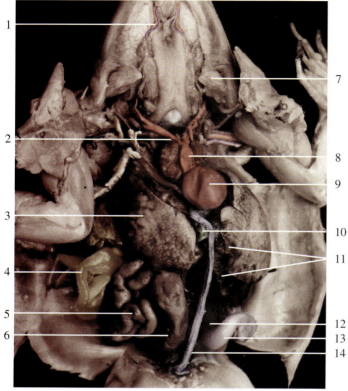

Figure 9.26 A ventral view of the frog viscera.

1. External carotid artery
2. Right aortic arch
3. Right lobe of liver
4. Fat body
5. Small intestine
6. Large intestine
7. Vocal sac (male only)
8. Truncus arteriosus
9. Heart
10. Gallbladder
11. Left lobe of liver
12. Stomach
13. Duodenum
14. Ventral abdominal vein

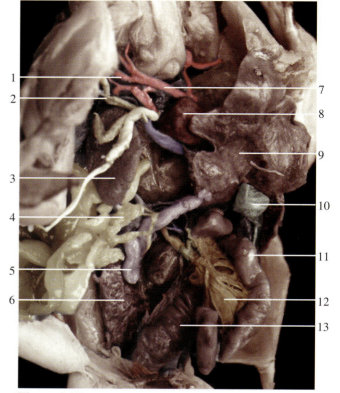

Figure 9.27 A deep view of the frog viscera.

1. Systemic arch
2. Pulmocutaneous arch
3. Right lung
4. Fat body
5. Testis
6. Right kidney
7. Common carotid artery
8. Heart
9. Liver
10. Gallbladder
11. Small intestine
12. Mesentery
13. Large Intestine

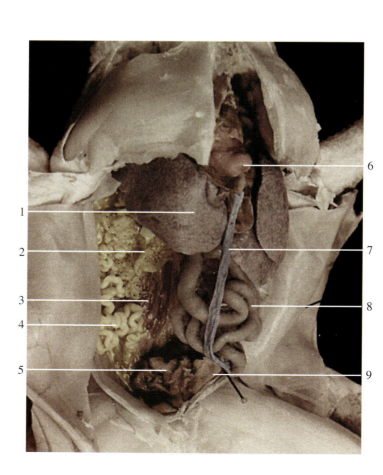

Figure 9.28 The female reproductive organs of a frog.

1. Liver
2. Ovary
3. Kidney
4. Oviduct
5. Uterus
6. Heart
7. Ventral abdominal vein
8. Small intestine
9. Urinary bladder (retracted)

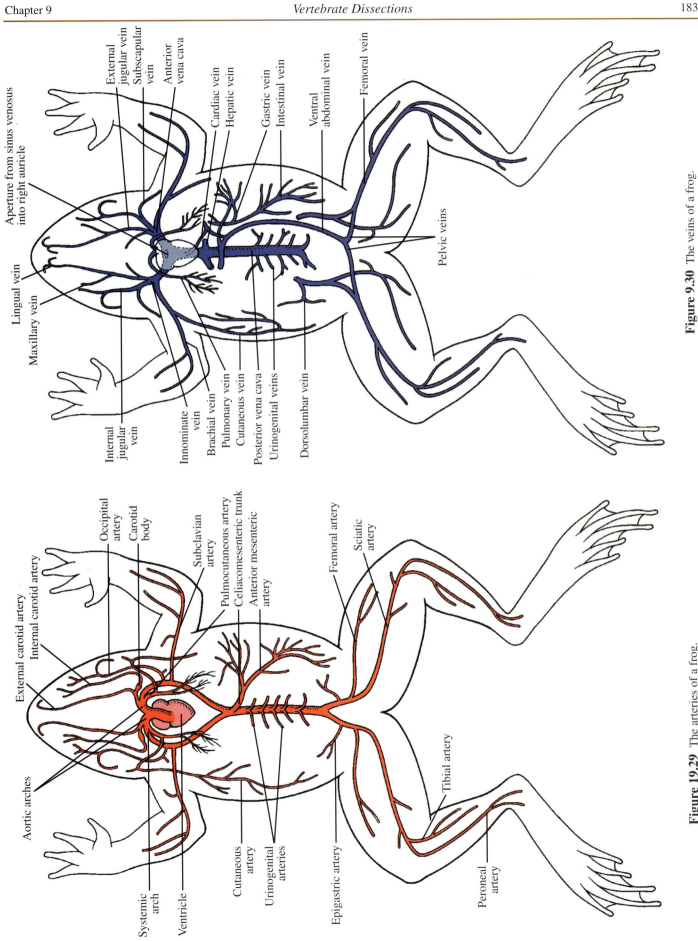

Figure 9.30 The veins of a frog.

Figure 19.29 The arteries of a frog.

Class Reptilia

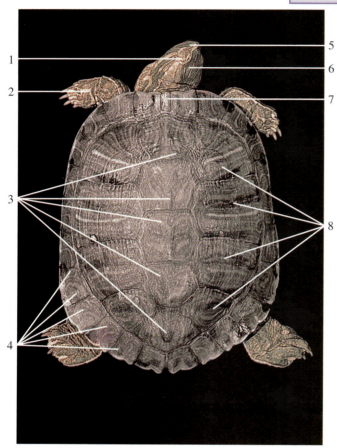

Figure 9.31 The dorsal view of a turtle carapace.

1. Eye
2. Pentadactyl foot
3. Vertebral scales
4. Marginal scales (encircle the carapace)
5. Nostril
6. Head
7. Nuchal scale
8. Costal scales

Figure 9.32 The ventral view of a turtle.

1. Gular scales
2. Pectoral scales
3. Femoral scales
4. Tail
5. Humeral scales
6. Abdominal scales
7. Anal scales

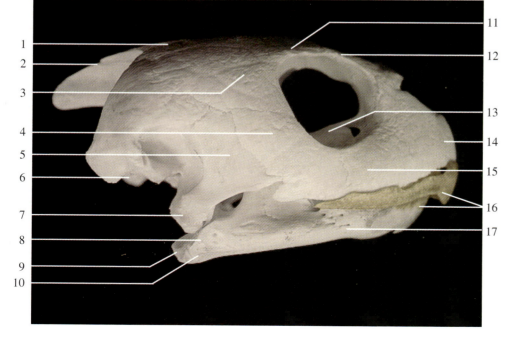

Figure 9.33 The skull of a turtle.

1. Parietal bone
2. Supraoccipital bone
3. Postorbital bone
4. Jugal bone
5. Quadratojugal bone
6. Exoccipital bone
7. Quadrate bone
8. Supraangular bone
9. Articular bone
10. Angular bone
11. Frontal bone
12. Prefrontal bone
13. Palatine bone
14. Premaxilla
15. Maxilla
16. Beak
17. Dentary

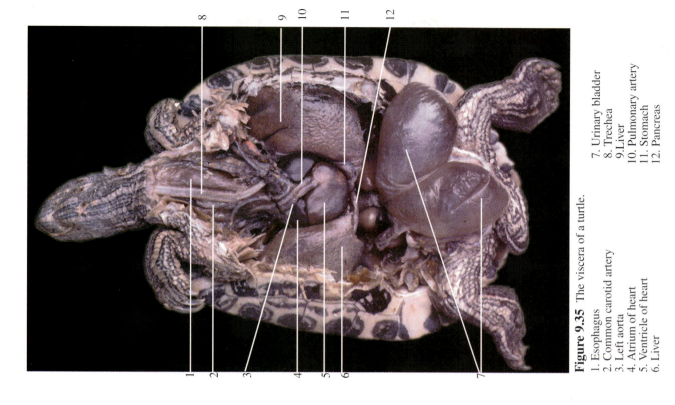

Figure 9.35 The viscera of a turtle.

1. Esophagus
2. Common carotid artery
3. Left aorta
4. Atrium of heart
5. Ventricle of heart
6. Liver

7. Urinary bladder
8. Trechea
9. Liver
10. Pulmonary artery
11. Stomach
12. Pancreas

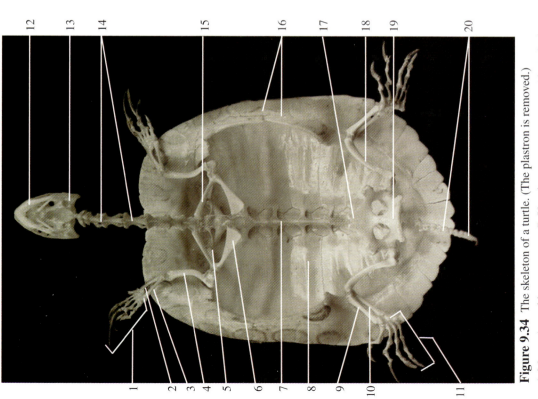

Figure 9.34 The skeleton of a turtle. (The plastron is removed.)

1. Manus (carpal bones, metacarpal bones, phalanges)
2. Radius
3. Ulna
4. Humerus
5. Procoracoid
6. Scapula

7. Vertebra
8. Rib
9. Tibia
10. Fibula
11. Pes (tarsal bones, metatarsal bones, phalanges)
12. Dentary

13. Articular
14. Cervical vertebrae
15. Acromion process
16. Dermal plate of carapace
17. Pubis
18. Femur
19. Ischium
20. Caudal vertebrae

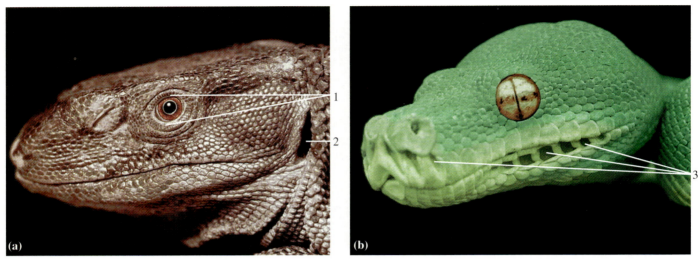

Figure 9.36 A comparison of the external anatomy of the head of (a) the savannah monitor lizard (*Varanus exanthematicus*) and (b) the green tree python (*Chondropython viridis*). Lizards have eyelids and external ears, whereas, these structures are lacking in snakes. (Notice the heat pit receptors, which are characteristic of pythons and pit vipers and are adaptive for predation on warm-blooded vertebrates.)

1. Eyelids   2. External ear   3. Heat pits

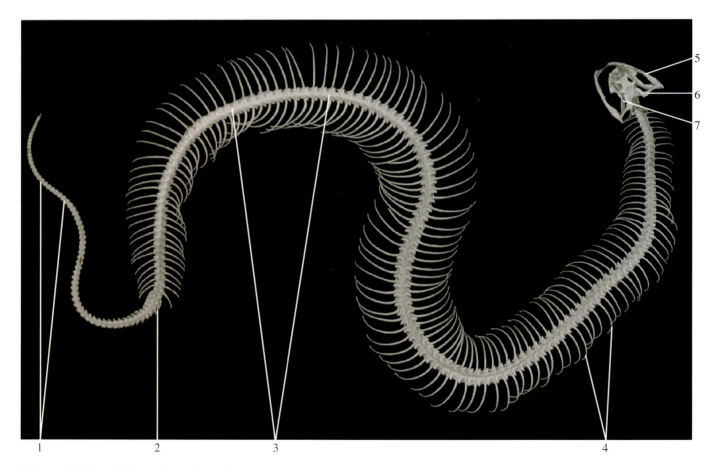

Figure 9.37 The skeleton of a snake (python).

1. Caudal vertebrae   5. Dentary
2. Vestigial pelvic girdle   6. Quadrate bone
3. Trunk vertebrae   7. Supratemporal bone
4. Ribs

Class Aves

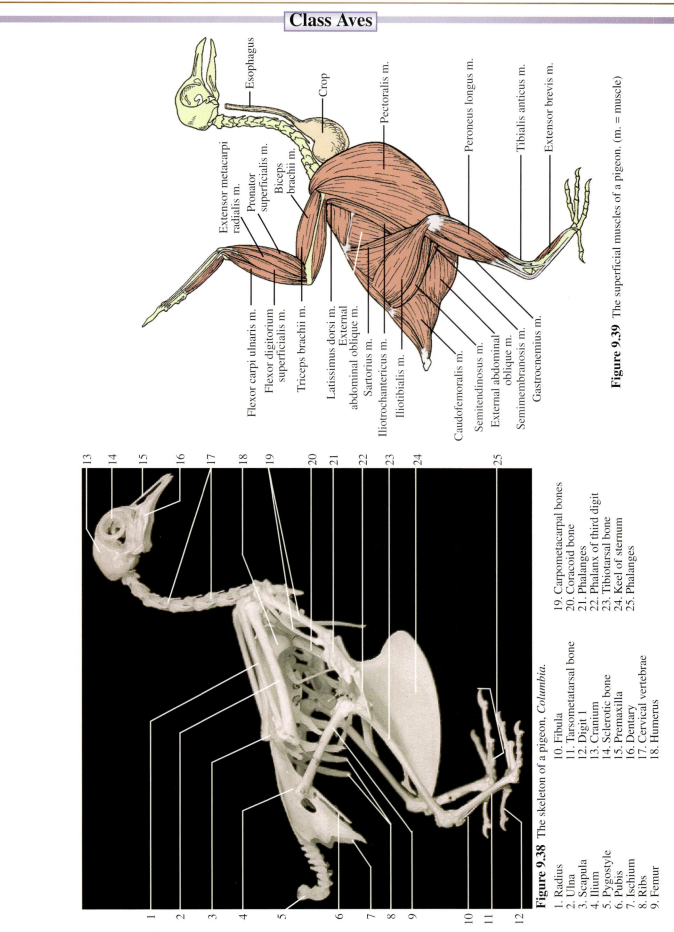

Figure 9.39 The superficial muscles of a pigeon. (m. = muscle)

Figure 9.38 The skeleton of a pigeon, *Columbia.*

1. Radius
2. Ulna
3. Scapula
4. Ilium
5. Pygostyle
6. Pubis
7. Ischium
8. Ribs
9. Femur
10. Fibula
11. Tarsometatarsal bone
12. Digit 1
13. Cranium
14. Sclerotic bone
15. Premaxilla
16. Dentary
17. Cervical vertebrae
18. Humerus
19. Carpometacarpal bones
20. Coracoid bone
21. Phalanges
22. Phalanx of third digit
23. Tibiotarsal bone
24. Keel of sternum
25. Phalanges

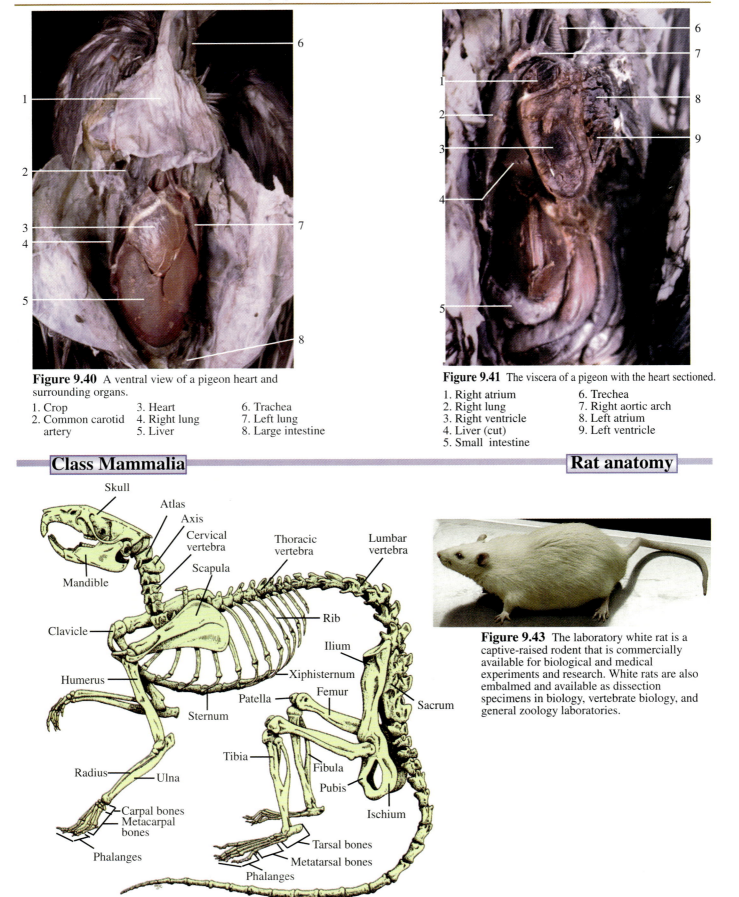

Figure 9.40 A ventral view of a pigeon heart and surrounding organs.

1. Crop	3. Heart	6. Trachea
2. Common carotid artery	4. Right lung	7. Left lung
	5. Liver	8. Large intestine

Figure 9.41 The viscera of a pigeon with the heart sectioned.

1. Right atrium	6. Trechea
2. Right lung	7. Right aortic arch
3. Right ventricle	8. Left atrium
4. Liver (cut)	9. Left ventricle
5. Small intestine	

Class Mammalia

Rat anatomy

Figure 9.42 The skeleton of a rat.

Figure 9.43 The laboratory white rat is a captive-raised rodent that is commercially available for biological and medical experiments and research. White rats are also embalmed and available as dissection specimens in biology, vertebrate biology, and general zoology laboratories.

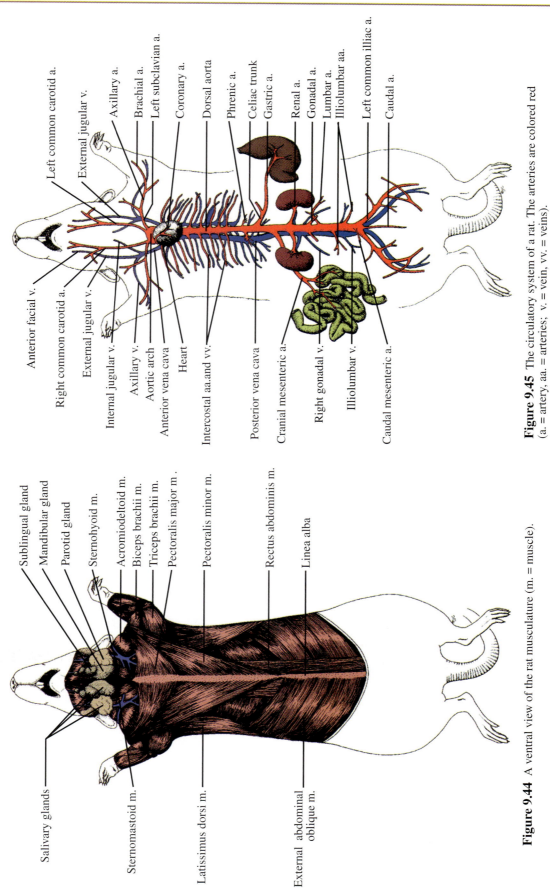

Figure 9.45 The circulatory system of a rat. The arteries are colored red (a. = artery, aa. = arteries; v. = vein, vv. = veins).

Figure 9.44 A ventral view of the rat musculature (m. = muscle).

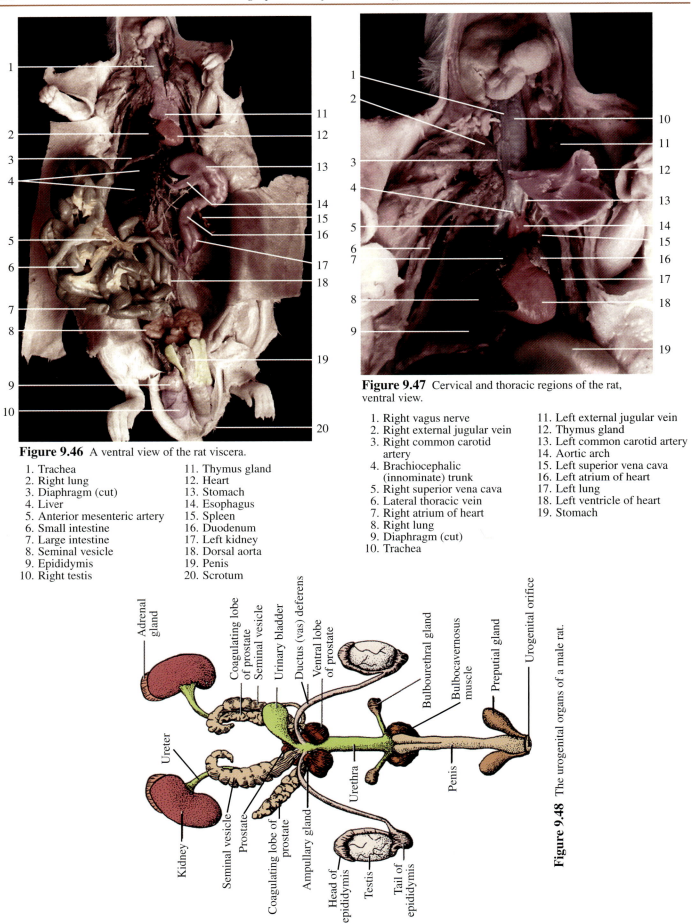

Figure 9.46 A ventral view of the rat viscera.

1. Trachea
2. Right lung
3. Diaphragm (cut)
4. Liver
5. Anterior mesenteric artery
6. Small intestine
7. Large intestine
8. Seminal vesicle
9. Epididymis
10. Right testis
11. Thymus gland
12. Heart
13. Stomach
14. Esophagus
15. Spleen
16. Duodenum
17. Left kidney
18. Dorsal aorta
19. Penis
20. Scrotum

Figure 9.47 Cervical and thoracic regions of the rat, ventral view.

1. Right vagus nerve
2. Right external jugular vein
3. Right common carotid artery
4. Brachiocephalic (innominate) trunk
5. Right superior vena cava
6. Lateral thoracic vein
7. Right atrium of heart
8. Right lung
9. Diaphragm (cut)
10. Trachea
11. Left external jugular vein
12. Thymus gland
13. Left common carotid artery
14. Aortic arch
15. Left superior vena cava
16. Left atrium of heart
17. Left lung
18. Left ventricle of heart
19. Stomach

Figure 9.48 The urogenital organs of a male rat.

Cat anatomy

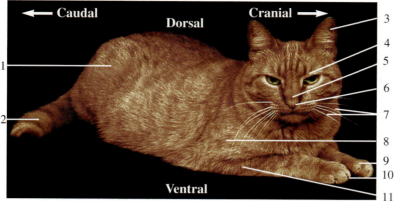

Figure 9.49 Directional terminology and superficial structures in a cat (quadrupedal vertebrate).

1. Thigh
2. Tail
3. Auricle (pinna)
4. Superior palpebra
    (superior eyelid)
5. Bridge of nose
6. Naris (nostril)
7. Vibrissae
8. Brachium
9. Manus (front foot)
10. Claw
11. Antebrachium

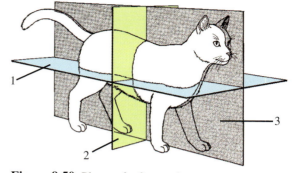

Figure 9.50 Planes of reference in a cat.

1. Coronal plane (frontal plane)
2. Transverse plane (cross-sectional plane)
3. Midsagittal plane (median plane)

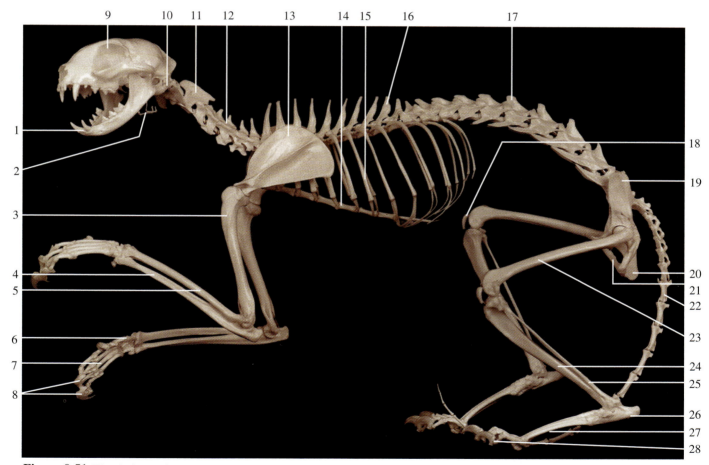

Figure 9.51 The skeleton of a cat.

1. Mandible
2. Hyoid bone
3. Humerus
4. Ulna
5. Radius
6. Carpal bones
7. Metacarpal bones
8. Phalanges
9. Skull
10. Atlas
11. Axis
12. Cervical vertebra
13. Scapula
14. Sternum
15. Rib
16. Thoracic vertebra
17. Lumbar vertebra
18. Patella
19. Ilium
20. Ischium
21. Pubis
22. Caudal vertebra
23. Femur
24. Tibia
25. Fibula
26. Tarsal bones
27. Metatarsal bones
28. Phalanges

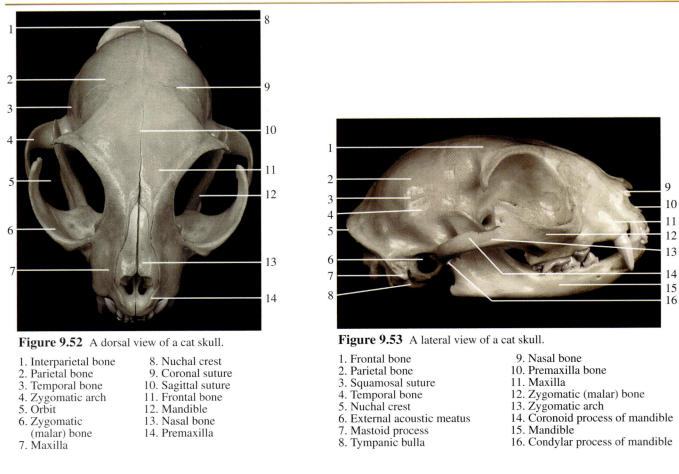

Figure 9.52 A dorsal view of a cat skull.

1. Interparietal bone
2. Parietal bone
3. Temporal bone
4. Zygomatic arch
5. Orbit
6. Zygomatic (malar) bone
7. Maxilla
8. Nuchal crest
9. Coronal suture
10. Sagittal suture
11. Frontal bone
12. Mandible
13. Nasal bone
14. Premaxilla

Figure 9.53 A lateral view of a cat skull.

1. Frontal bone
2. Parietal bone
3. Squamosal suture
4. Temporal bone
5. Nuchal crest
6. External acoustic meatus
7. Mastoid process
8. Tympanic bulla
9. Nasal bone
10. Premaxilla bone
11. Maxilla
12. Zygomatic (malar) bone
13. Zygomatic arch
14. Coronoid process of mandible
15. Mandible
16. Condylar process of mandible

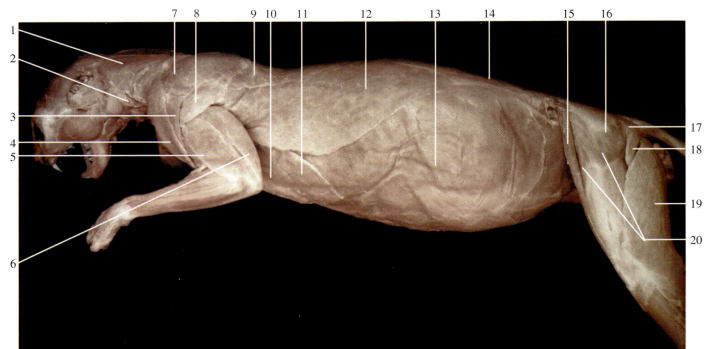

Figure 9.54 A lateral view of the superficial muscles of the cat.

1. Clavotrapezius m.
2. Sternomastoid m.
3. Acromiodeltoid m.
4. Clavobrachialis m.
5. Lateral head of triceps brachii m.
6. Long head of triceps brachii m.
7. Acromiotrapezius m.
8. Spinodeltoid m.
9. Spinotrapezius m.
10. Pectoralis minor m.
11. Xiphihumeralis m.
12. Latissimus dorsi m.
13. External abdominal oblique m.
14. Lumbodorsal fascia
15. Sartorius m.
16. Gluteus medius m.
17. Gluteus maximus m.
18. Caudofemoralis m.
19. Biceps femoris m.
20. Tensor fasciae latae m.

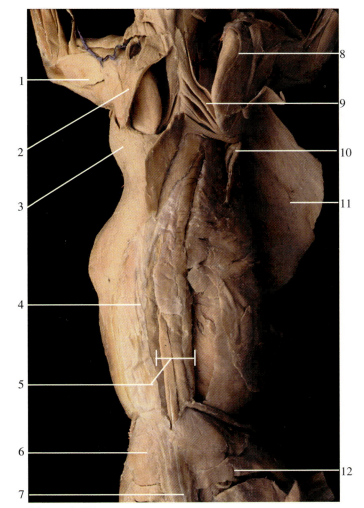

Figure 9.55 A dorsal view of the superficial muscles of the cat.

1. Lateral head of
   triceps brachii m.
2. Acromiotrapezius m.
3. Latissimus dorsi m.
4. Lumbodorsal fascia
5. Sacrospinalis m.
6. Gluteus medius m.
7. Caudal m..
8. Supraspinatus m
9. Rhomboideus m.
10. Serratus anterior m.
11. Latissimus dorsi m.
12. Gluteus maximus m.

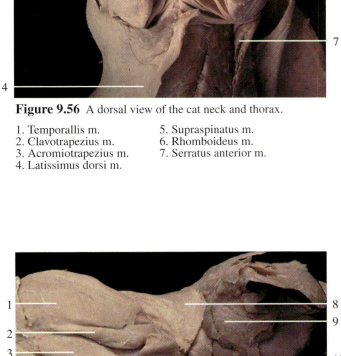

Figure 9.56 A dorsal view of the cat neck and thorax.

1. Temporallis m.
2. Clavotrapezius m.
3. Acromiotrapezius m.
4. Latissimus dorsi m.
5. Supraspinatus m.
6. Rhomboideus m.
7. Serratus anterior m.

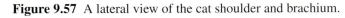

Figure 9.57 A lateral view of the cat shoulder and brachium.

1. Acromiotrapezius m.
2. Levator scapulae ventralis m.
3. Spinodeltoid m.
4. Latissimus dorsi m.
5. Long head of triceps brachii m.
6. Clavobrachialis m.
7. Lateral head of triceps
   brachii m.
8. Clavotrapezius m.
9. Auricular m.
10. Acromiodeltoid m.

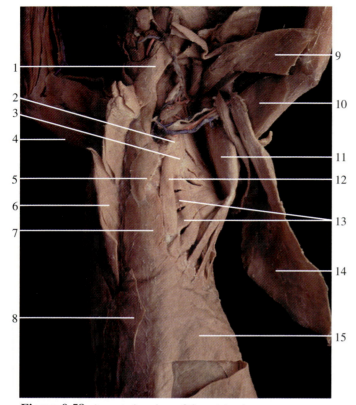

Figure 9.58 An anterior view of the cat trunk.

1. Sternomastoid m.
2. Scalenus anterior m.
3. Scalenus posterior m.
4. Epitrochlearis m.
5. Transverse costarum m.
6. Pectoralis minor m. (cut)
7. Rectus abdominis m.
8. Xiphihumeralis m. (cut)
9. Pectoralis minor (cut)
10. Epitrochlearis m.
11. Subscapularis m.
12. Scalenus medius m.
13. Serratus anterior m.
14. Latissimus dorsi m. (cut)
15. External abdominal oblique m.

Figure 9.59 A lateral view of the cat trunk.

1. External abdominal oblique m.
2. Latissimus dorsi m.
3. Spinodeltoid m.
4. Transverse abdominis m.
5. Long head of triceps brachii m.
6. Serratus anterior m.
7. Internal abdominal oblique m.
8. Tensor fascia latae
9. Caudofemoris m.
10. Vastus lateralis m.
11. Sartorius m.

Figure 9.61 A medial view of the cat thigh and leg.

1. Rectus abdominus m.
2. Adductor longus m.
3. Adductor femoris m.
4. Semimembranosus m.
5. Gracilis m. (cut)
6. Tendo calcaneus (Achilles tendon)
7. Sartorius m.
8. Tensor fascia latae m.
9. Vastus lateralis m.
10. Rectus femoris m.
11. Vastus medialis m.
12. Flexor digitorum longus m.
13. Tibialis anterior m.

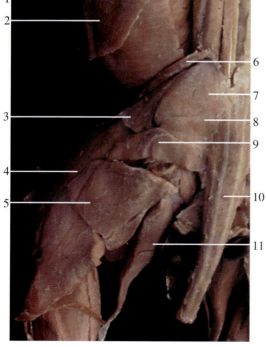

Figure 9.60 A lateral view of the superficial cat thigh.

1. Internal abdominal oblique m.
2. External abdominal oblique m.
3. Tensor fascia latae (cut)
4. Vastus lateralis m.
5. Biceps femoris m.
6. Sartorius m.
7. Gluteus medius m.
8. Gluteus maximus m.
9. Caudofemoris m.
10. Caudal m.
11. Semitendinosus m.

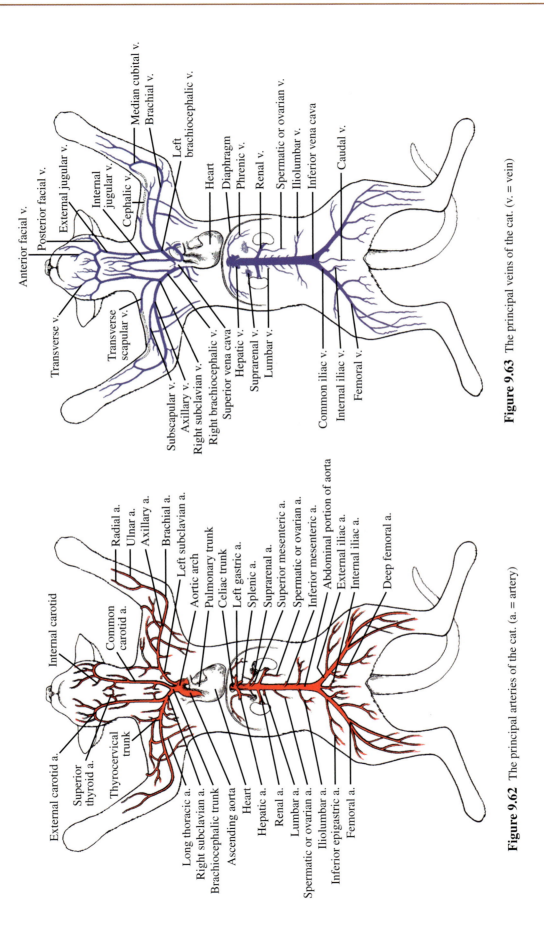

Figure 9.63 The principal veins of the cat. (v. = vein)

Figure 9.62 The principal arteries of the cat. (a. = artery)

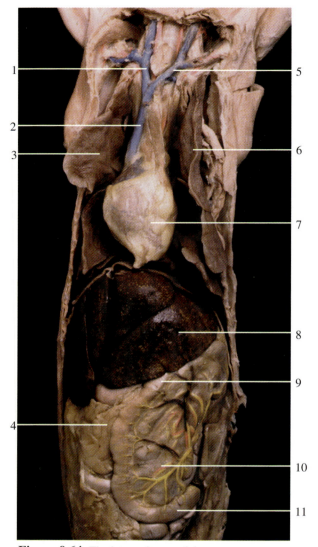

Figure 9.64 The intact viscera of the cat.

1. Right brachiocephalic vein	7. Heart
2. Superior vena cava	8. Liver
3. Right lung	9. Stomach
4. Greater omentum	10. Mesentery
5. Left brachiocephalic vein	11. Small intestine
6. Left lung	

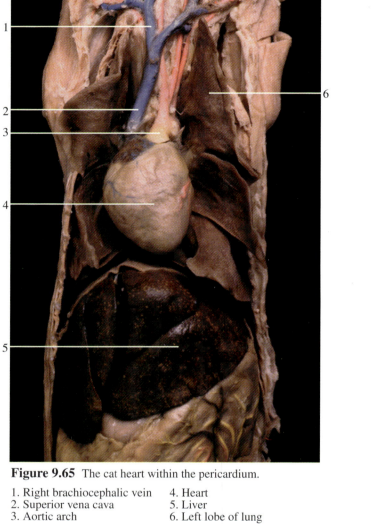

Figure 9.65 The cat heart within the pericardium.

1. Right brachiocephalic vein	4. Heart
2. Superior vena cava	5. Liver
3. Aortic arch	6. Left lobe of lung

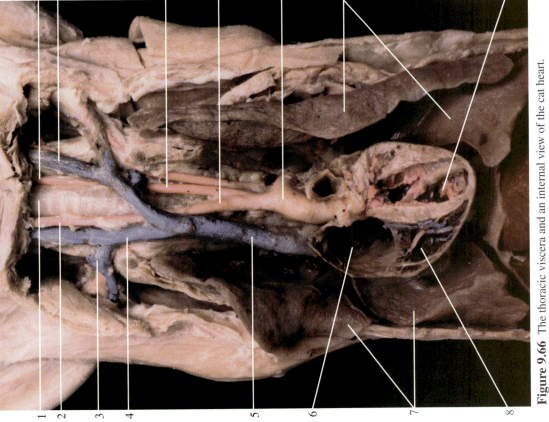

Figure 9.67 The cat heart and surrounding structures.

1. Trachea
2. Common carotid arteries
3. Axillary vein
4. Heart (cut)
5. Left ventricle
6. Vagus nerve
7. External jugular vein

8. Left subclavian vein
9. Superior vena cava
10. Brachiocephalic trunk
11. Thoracic aorta

Figure 9.66 The thoracic viscera and an internal view of the cat heart.

1. Trachea
2. Right common carotid artery
3. Right subclavian vein
4. Right brachiocephalic vein
5. Superior vena cava
6. Right atrium
7. Right lung
8. Right ventricle

9. Left common carotid artery
10. Internal jugular vein
11. Brachiocephalic artery
12. Right common carotid artery
13. Aortic arch
14. Left lung
15. Left ventricle

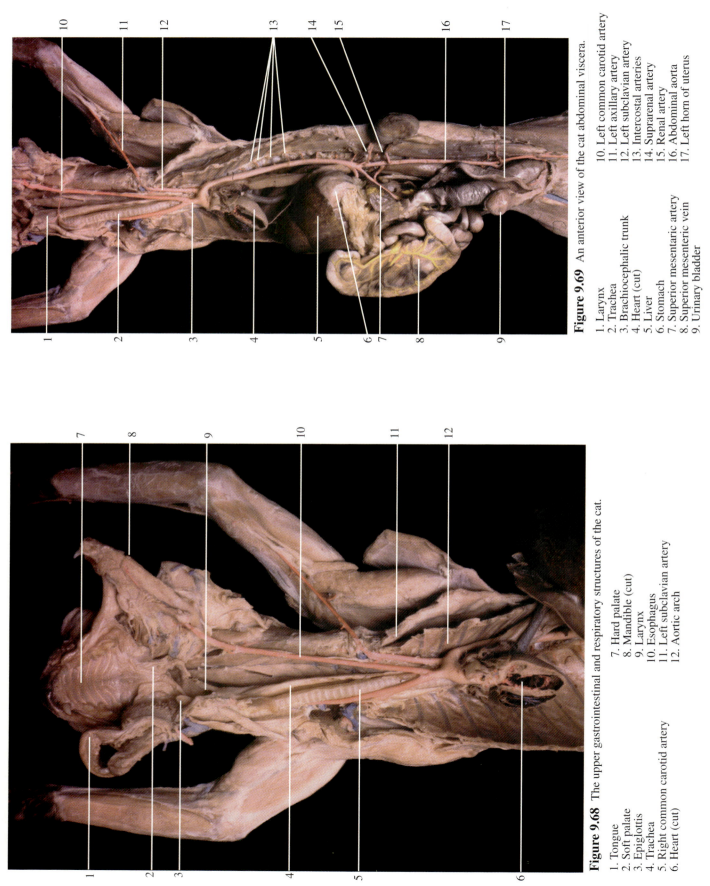

Figure 9.69 An anterior view of the cat abdominal viscera.

1. Larynx
2. Trachea
3. Brachiocephalic trunk
4. Heart (cut)
5. Liver
6. Stomach
7. Superior mesenteric artery
8. Superior mesenteric vein
9. Urinary bladder

10. Left common carotid artery
11. Left axillary artery
12. Left subclavian artery
13. Intercostal arteries
14. Suprarenal artery
15. Renal artery
16. Abdominal aorta
17. Left horn of uterus

Figure 9.68 The upper gastrointestinal and respiratory structures of the cat.

1. Tongue
2. Soft palate
3. Epiglottis
4. Trachea
5. Right common carotid artery
6. Heart (cut)

7. Hard palate
8. Mandible (cut)
9. Larynx
10. Esophagus
11. Left subclavian artery
12. Aortic arch

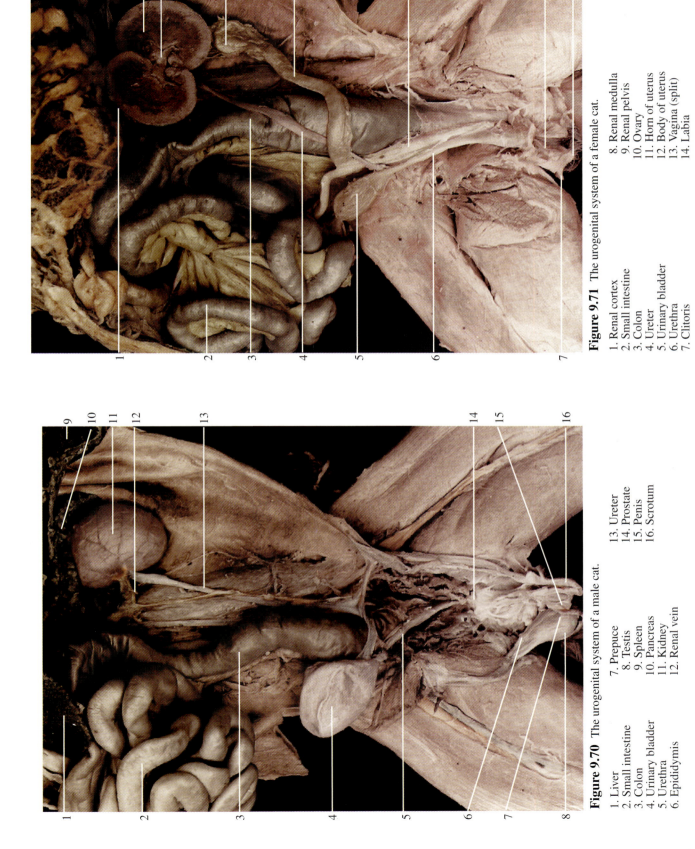

Figure 9.71 The urogenital system of a female cat.

1. Renal cortex
2. Small intestine
3. Colon
4. Ureter
5. Urinary bladder
6. Urethra
7. Clitoris
8. Renal medulla
9. Renal pelvis
10. Ovary
11. Horn of uterus
12. Body of uterus
13. Vagina (split)
14. Labia

Figure 9.70 The urogenital system of a male cat.

1. Liver
2. Small intestine
3. Colon
4. Urinary bladder
5. Urethra
6. Epididymis
7. Prepuce
8. Testis
9. Spleen
10. Pancreas
11. Kidney
12. Renal vein
13. Ureter
14. Prostate
15. Penis
16. Scrotum

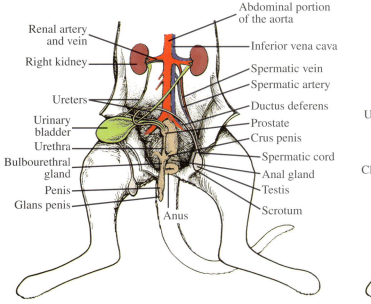

Renal artery and vein
Right kidney
Ureters
Urinary bladder
Urethra
Bulbourethral gland
Penis
Glans penis

Abdominal portion of the aorta
Inferior vena cava
Spermatic vein
Spermatic artery
Ductus deferens
Prostate
Crus penis
Spermatic cord
Anal gland
Testis
Scrotum

Anus

Figure 9.72 A diagram of the urogenital system of a male cat.

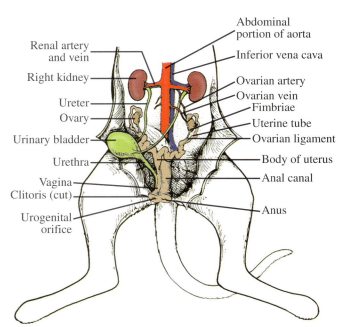

Renal artery and vein
Right kidney
Ureter
Ovary
Urinary bladder
Urethra
Vagina
Clitoris (cut)
Urogenital orifice

Abdominal portion of aorta
Inferior vena cava
Ovarian artery
Ovarian vein
Fimbriae
Uterine tube
Ovarian ligament
Body of uterus
Anal canal
Anus

Figure 9.73 A diagram of the urogenital system of a female cat.

Fetal pig anatomy

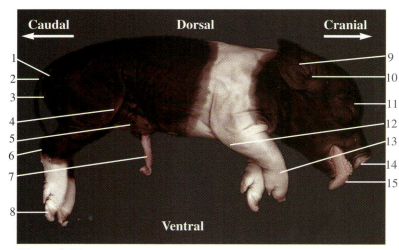

Caudal   Dorsal   Cranial

Ventral

Figure 9.74 Directional terminology and superficial structures in a fetal pig (quadrupedal vertebrate).

1. Anus
2. Tail
3. Scrotum
4. Knee
5. Teat
6. Ankle
7. Umbilical cord
8. Hoof
9. Auricle (pinna)
10. External auditory canal
11. Superior palpebra (superior eyelid)
12. Elbow
13. Wrist
14. Naris (nostril)
15. Tongue

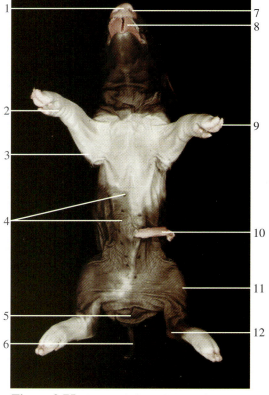

Figure 9.75 A ventral view of the surface anatomy of the fetal pig.

1. Nose
2. Wrist
3. Elbow
4. Teats
5. Scrotum
6. Tail
7. Nostril
8. Tongue
9. Digit
10. Umbilical cord
11. Knee
12. Ankle

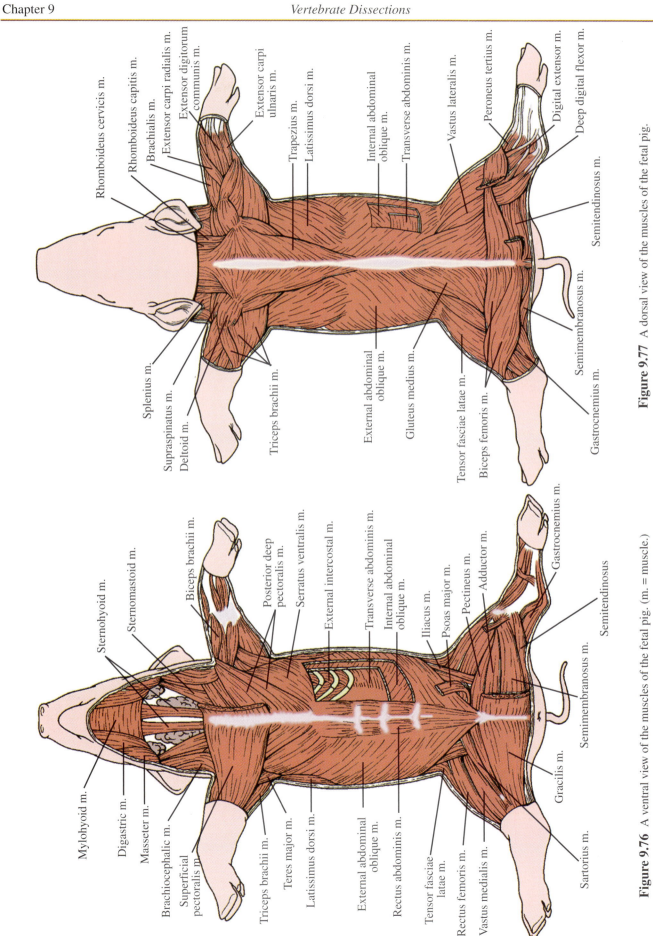

Figure 9.77 A dorsal view of the muscles of the fetal pig.

Figure 9.76 A ventral view of the muscles of the fetal pig. (m. = muscle.)

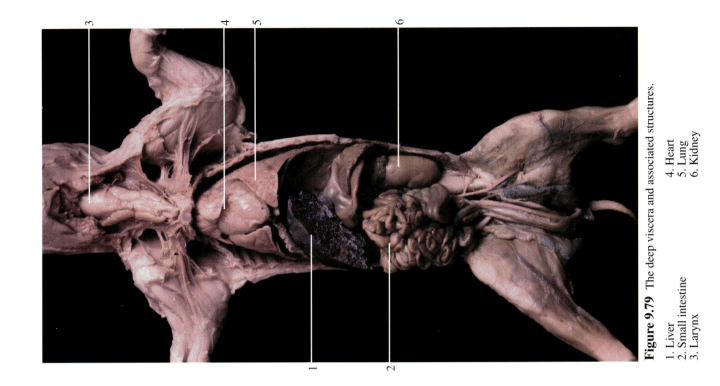

Figure 9.79 The deep viscera and associated structures.

1. Liver
2. Small intestine
3. Larynx

4. Heart
5. Lung
6. Kidney

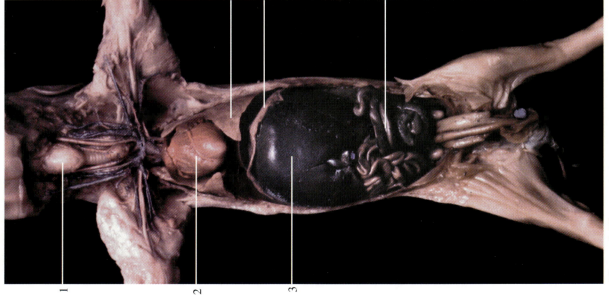

Figure 9.78 A ventral view of the viscera of the fetal pig.

1. Larynx
2. Heart
3. Liver
4. Lung

5. Diaphragm
6. Small intestine

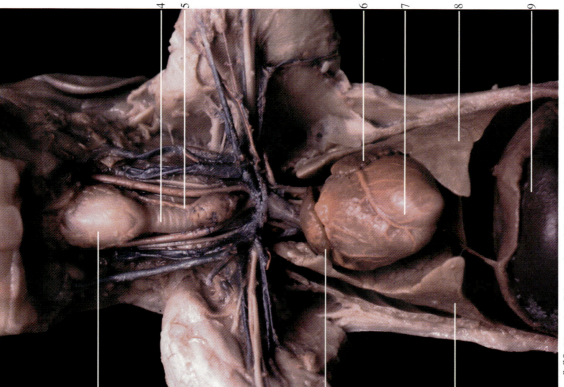

Figure 8.81 Vessels of the abdomen and the lower extremities of the fetal pig.

1. Heart
2. Thoracic aorta
3. Kidney
4. Renal vein
5. Ductus deferens

6. Small intestine
7. Colon

Figure 9.80 Vessels of the neck and the thoracic organs of the fetal pig.

1. Larynx
2. Right auricle
3. Right lung
4. Trachea
5. Common carotid artery

6. Left auricle
7. Heart
8. Left lung
9. Diaphragm

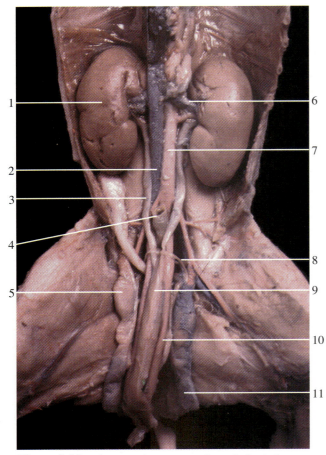

Figure 9.82 The urogenital system of a male fetal pig.

1. Kidney
2. Inferior vena cava
3. Ureter
4. Rectum (cut)
5. Partially dissected testis
6. Renal vein
7. Descending aorta
8. Ductus deferens
9. Urinary bladder
10. Umbilical vein
11. Epididymis

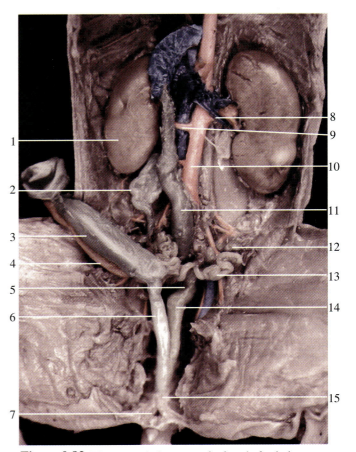

Figure 9.83 The urogenital system of a female fetal pig.

1. Right kidney
2. Ureter
3. Urinary bladder
4. Umbilical artery
5. Uterus
6. Urethra
7. Urogenital orifice
8. Renal vein
9. Renal artery
10. Abdominal descending
 aorta
11. Rectum
12. Ovary
13. Horn of uterus
14. Vaginal
15. Urogenital sinus

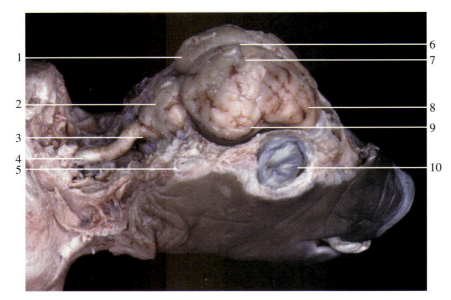

Figure 9.84 General structures of the fetal pig brain. Because the cerebrum is less defined in pigs, the regions are not known as lobes as they are in humans.

1. Occipital region of cerebrum
2. Cerebellum
3. Medulla oblongata
4. Spinal cord
5. External acoustic meatus
6. Longitudinal fissure
7. Parietal region of cerebrum
8. Frontal region of cerebrum
9. Temporal region of cerebrum
10. Eye

Sheep heart dissection

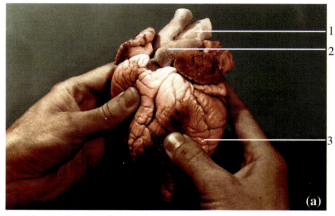

Position the heart so the ventral surface faces you. Notice the thicker ventricular walls, especially the left ventricle.

1. Aortic arch
2. Pulmonary trunk
3. Left ventricle

Insert the scissors into the superior vena cava. The cut should expose the interior of the right atrium. Notice the right tricuspid (atrioventricular) valve.

1. Right atrium
2. Superior vena cava

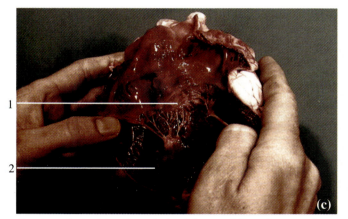

Continue the incision through the right ventricle to the apex of the heart. Observe the structure of the valve.

1. Right tricuspid valve
2. Right ventricle

Begin the next incision in the left atrium. This time, continue through both the atrium and the ventricle.

1. Left ventricle
2. Left atrium

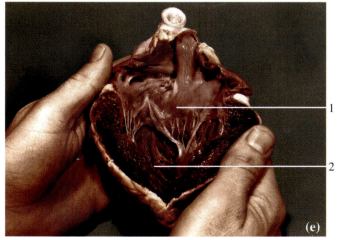

Expose the left ventricle and atrium. Notice the difference between the right and left ventricles, especially the thicker muscular wall of the left ventricle.

Figure 9.85 Sheep heart dissections.

1. Left bicuspid (mitral) valve.
2. Left ventricle

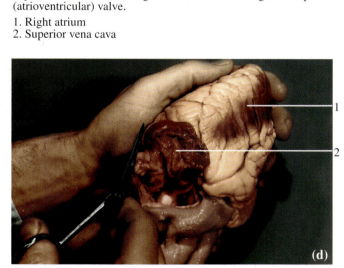

Appendix – Glossary of Terms

abdomen – the portion of the trunk of the mammalian body located between the diaphragm and the pelvis, that contains the abdominal cavity and its visceral organs; one of the three principal body regions (head, thorax, and abdomen) of many animals.

abduction – a movement away from the axis or midline of the body; opposite of adduction, a movement of a digit away from the axis of a limb.

abiotic – without living organisms; nonliving portions of the environment.

abscission – the shedding of leaves, flowers, fruits, or other plant parts, usually following the formation of an abscission zone.

absorption – movement of a substance into a cell or an organism, or through a surface within an organism.

acapnia – a decrease in normal amount of CO_2 in the blood.

accommodation – a change in the shape of the lens of the eye so that vision is more acute; the focusing for various distances.

acetone – an organic compound that may be present in the urine of diabetics; also called *ketone body*.

acetylcholine – a neurotransmitter chemical secreted at the terminal ends of many neurons, responsible for postsynaptic transmission; also called *ACh*.

acetylcholinesterase – an enzyme that breaks down acetylcholine; also called *AChE*.

Achilles tendon – see *tendo calcaneus*.

acid – a substance that releases hydrogen ions (H^+) in a solution.

acidosis – a disorder of body chemistry in which the alkaline substances of the blood are reduced in amount below normal.

acoelomate – without a coelomic cavity; as in flatworms.

acoustic – referring to sound or the sense of hearing.

actin – a protein in muscle fibers that together with myosin is responsible for contraction.

action potential – the change in ionic charge propagated along the membrane of a neuron; the nerve impulse.

active transport – movement of a substance into or out of a cell from a lesser to a greater concentration, requiring a carrier molecule and expenditure of energy.

adaptation –structural, physiological, or behavioral traits of an organism that promote its survival and contribute to its ability to reproduce under specific environmental conditions.

adduction – a movement toward the axis or midline of the body; opposite of abduction, a movement of a digit toward the axis of a limb.

adenohypophysis – anterior pituitary.

adenoid – paired lymphoid structures in the nasopharynx; also called *pharyngeal tonsils*.

adenosine triphosphate (ATP) – a chemical compound that provides energy for cellular use.

adhesion – the attraction between unlike substances.

adipose – fat, or fat-containing, such as adipose tissue.

adrenal glands – endocrine glands; one superior to each kidney; also called *suprarenal glands*.

adventitious root – supportive root developing from the stem of a plant.

aerobic – requiring free O_2 for growth and metabolism as in the case of certain bacteria called aerobes.

agglutination – clumping of cells; particular reference to red blood cells in an antigen-antibody reaction.

aggregate fruit – ripened ovaries from a single flower with several separate carpels.

aggression – provoking, domineering behavior.

alga (pl. *algae*) – any of a diverse group of aquatic photosynthesizing organisms that are either unicellular or are multicellular; algae comprise the phytoplankton and seaweeds of the Earth.

alkaline – a substance having a pH greater than 7.0; basic.

allantois – an extraembryonic membranous sac that forms blood cells and gives rise to the fetal umbilical arteries and vein. It also contributes to the formation of the urinary bladder.

allele – an alternative form of a gene occurring at a given chromosome site, or locus.

all-or-none response – functioning completely when exposed to a stimulus of threshold strength; applies to action potentials through neurons and muscle fiber contraction.

alpha helix – right-handed spiral typical in proteins and DNA.

alternation of generations – two-phased life cycle characteristic of many plants in which there are sporophyte and gametophyte generations.

altruism – behavior benefiting other organisms without regard to its possible advantage or detrimental effect on the performer.

alveolus – an individual air capsule within the lung. Alveoli are the basic functional units of respiration. Also, the socket that secures a tooth.

amino acid – a unit of protein that contains an amino group (NH_2) and an acid group (COOH).

amnion – a membrane that surrounds the fetus to contain the amniotic fluid.

amniote – an animal that has an amnion during embryonic development; reptiles, birds, and mammals.

amoeba – protozoans that move by means of pseudopodia.

amphiarthrosis – a slightly moveable joint in a functional classification of joints.

anaerobic respiration – metabolizing and growing in the absence of oxygen.

analogous – similar in function regardless of developmental origin; generally in reference to similar adaptations.

anatomical position – the position in human anatomy in which there is an erect body stance with the eyes directed forward, the arms at the sides, and the palms of the hands facing forward.

anatomy – the branch of science concerned with the structure of the body and the relationship of its organs.

angiosperm – flowering plant, having double fertilization resulting in development of specialized seeds within fruits.

annual – a flowering plant that completes its entire life cycle in a single year or growing season.

annual rings – yearly growth demarcations in woody plants formed by buildup of secondary xylem.

annulus – a ringlike segment, such as body rings on leeches.

antebrachium – the forearm.

antenna – a sensory appendage on many species of invertebrate animals.

anterior (ventral) – toward the front; the opposite of posterior (dorsal).

anther – the portion of a plant stamen in which pollen is produced.

antheridium – male reproductive organ in certain nonseed plants and algae where motile sperm are produced.

anticodon – three ("a triplet") nucleotide sequence in transfer RNA that pairs with a complementary codon (triplet) in messenger RNA.

antigen – a foreign material, usually a protein, that triggers the immune system to produce antibodies.

anus – the terminal end of the GI tract, opening of the anal canal.

aorta – the major systemic vessel of the arterial portion of the circulatory system, emerging from the left ventricle.

apical meristem – embryonic plant tissue in the tip of a root, bud, or shoot where continual cell divisions cause growth in length.

apocrine gland – a type of sweat gland that functions in evaporative cooling.

apopyle – opening of the radial canal into the spongocoel of sponges.

appeasement – submissive behavior, usually soliciting an end of aggression.

appendix – a short pouch that attaches to the cecum.

aqueous humor – the watery fluid that fills the anterior and posterior chambers of the eye.

arachnoid mater – the weblike middle covering (meninx) of the central nervous system.

arbor vitae – the branching arrangement of white matter within the cerebellum.

archaebacteria – organisms within the kingdom Monera that represent an early group of simple life forms.

archegonium – female reproductive organ in certain nonseed plants and algae where eggs are produced.

archenteron – the principal cavity of an embryo during the gastrula stage. Lined with endoderm, the archenteron develops into the digestive tract.

areola – the pigmented ring around the nipple.

artery – a blood vessel that carries blood away from the heart.

articular cartilage – a hyaline cartilaginous covering over the articulating surface of bones of synovial joints.

ascending colon – the portion of the large intestine between the cecum and the hepatic (right colic) flexure.

asexual – lacking distinct sexual organs and lacking the ability to produce gametes.

aster – minute rays of microtubules at the ends of the spindle apparatus in animal cells during cell division.

asymmetry – not symmetrical.

atom – the smallest unit of an element that can exist and still have the properties of the element; collectively, atoms form molecules in a compound.

atomic number – the weight of the atom of a particular element.

atomic weight – the number of protons together with the number of neutrons within the nucleus of an atom.

ATP (adenosine triphosphate) – a compound of adenine, ribose, and three phosphates; it is the energy carrier for most cellular processes.

atrium – either of two superior chambers of the heart that receive venous blood.

atrophy – a wasting away or decrease in size of a cell or organ.

auditory tube – a narrow canal that connects the middle ear chamber to the pharynx; also called the *eustachian canal*.

autonomic – self-governing; pertaining to the division of the nervous system which controls involuntary activities.

autosome – a chromosome other than a sex chromosome.

autotroph – an organism capable of synthesizing its own organic molecules (food) from inorganic molecules.

axilla – the depressed hollow under the arm; the armpit.

axillary bud – a group of meristematic cells at the junction of a leaf and stem which develops branches or flowers; also called *lateral bud*.

axon – The elongated process of a neuron (nerve cell) that transmits an impulse away from the cell body.

bacillus (pl. *bacilli*) – a rod-shaped bacterium.

bacteria – prokaryotes within the kingdom Monera, lacking the organelles of eukaryotic cells.

bark – outer tissue layers of a tree consisting of cork, cork cambium, cortex, and phloem.

basal – at or near the base or point of attachment, as of a plant shoot.

base – a substance that contributes or liberates hydroxide ions in a solution.

basement membrane – a thin sheet of extracellular substance to which the basal surfaces of membranous epithelial cells are attached.

basidia – club-shaped reproductive structures of club fungi that produce basidiospores during sexual reproduction.

basophil – a granular leukocyte that readily stains with basophilic dye.

belly –the thickest circumference of a skeletal muscle.

benign – nonmalignant; a confined tumor.

berry – a simple fleshy fruit.

biennial – a plant that lives through two growing seasons; generally, these plants have only vegetative growth during the first season, and flower and set seed during the second.

bilateral symmetry – the morphologic condition of having similar right and left halves.

binary fission – a process of sexual reproduction that does not involve a mitotic spindle.

binomial nomenclature – assignment of two names to an organism, the first of which is the genus and the second the species—together, constituting the scientific name.

biome – a major climax community characterized by a particular group of plants and animals.

biosphere – the portion of the earth's atmosphere and surface where living organisms exist.

biotic – pertaining to aspects of life, especially to characteristics of ecosystems.

bisexual flower – a flower that contains both male and female sexual structures.

blade – the broad expanded portion of a leaf.

blastocoel – the cavity of a blastocyst.

blastula – an early stage of prenatal development between the morula and embryonic stages.

blood – the fluid connective tissue that circulates through the cardiovascular system to transport substances throughout the body.

bolus – a moistened mass of food that is swallowed from the oral cavity into the pharynx.

bone – an organ composed of solid, rigid connective tissue, forming a component of the skeletal system.

Bowman's capsule – see *glomerular capsule*.

brain – the enlarged superior portion of the central nervous system, located in the cranial cavity of the skull.

brain stem – the portion of the brain consisting of the medulla oblongata, pons, and midbrain.

bronchial tree – the bronchi and their branching bronchioles.

bronchiole – a small division of a bronchus within the lung.

bronchus – a branch of the trachea that leads to a lung.

budding – a type of asexual reproduction in which outgrowths from the parent plant pinch off to live independently or else remain attached forming colonies.

buccal cavity – the mouth, or oral cavity.

buffer – a compound or substance that prevents large changes in the pH of a solution.

bulb – a thickened underground stem often enclosed by enlarged, fleshy leaves containing stored food.

bursa – a saclike structure filled with synovial fluid, which occurs around joints.

buttock – the rump or fleshy mass on the posterior aspect of the lower trunk, formed primarily by the gluteal muscles.

calorie – the heat required to raise one kilogram of water one degree centigrade.

calyx – a cup-shaped portion of the renal pelvis that encircles renal papillae; the collective term for the sepals of a flower.

cambium – the layer of meristematic tissue in roots and stems of many vascular plants that continues to produce tissue.

cancellous bone – spongy bone; bone tissue with a latticelike structure.

capillary – a microscopic blood vessel that connects an arteriole and a venule; the functional unit of the circulatory system.

carapace – protective covering over the dorsal part of the body of certain crustaceans and turtles.

carcinogenic – stimulating or causing the growth of a malignant tumor, or cancer.

carnivore – any animal that feeds upon another; especially, flesh-eating mammal.

carpus – the proximal portion of the hand that contains the carpal bones.

carrying capacity – the maximum number of organisms of a species that can be maintained indefinitely in an ecosystem.

cartilage – a type of connective tissue with a solid elastic matrix.

catalyst – a chemical, such as an enzyme, that accelerates the rate of a reaction of a chemical process but is not used up in the process.

caudal – referring to a position more toward the tail.

cecum – the pouchlike portion of the large intestine to which the ileum of the small intestine is attached.

cell – the structural and functional unit of an organism; the smallest structure capable of performing all the functions necessary for life.

cell wall – a rigid protective structure of a plant cell surrounding the cell membrane; often composed of cellulose fibers embedded in a polysaccharide/protein matrix.

cellular respiration – the reactions of glycolysis, Krebs cycle, and electron transport system that provide cellular energy and accompanying reactions to produce ATP.

cellulose – a polysaccharide produced as fibers that forms a major part of the rigid cell wall around a plant cell.

central nervous system (CNS) – the brain and the spinal cord.

centromere – a portion of the chromosome to which a spindle fiber attaches during mitosis or meiosis.

centrosome – a dense body near the nucleus of a cell that contains a pair of centrioles.

cephalothorax – fusion of the head and thoracic regions characteristic of certain arthropods.

cercaria – larva of trematodes (flukes).

cerebellum – the portion of the brain concerned with the coordination of movements and equilibrium.

cerebrospinal fluid – a fluid that buoys and cushions the central nervous system.

cerebrum – the largest portion of the brain, composed of the right and left hemispheres.

cervical – pertaining to the neck or a necklike portion of an organ.

chelipeds – front pair of pincerlike legs in most decapod crustaceans, adapted for seizing and crushing.

chitin – strong, flexible polysaccharide forming the exoskeleton of arthropods.

chlorophyll – green pigment in photosynthesizing organisms that absorbs energy from the sun's rays.

chloroplast – a membrane-enclosed organelle which contains chlorophyll and is the site of photosynthesis.

choanae – the two posterior openings from the nasal cavity into the nasopharynx.

cholesterol – a lipid used in the synthesis of steroid hormones.

chondrocyte – a cartilage cell.

chorion – an extraembryonic membrane that participates in the formation of the placenta.

choroid – the vascular, pigmented middle layer of the wall of the eye.

chromatin – threadlike network of DNA and proteins within the nucleus.

chromosome – structure in the nucleus that contains the genes for genetic expression.

chyme – the mass of partially digested food that passes from the stomach into the duodenum of the small intestine.

cilia – microscopic, hairlike processes that move in a wavelike manner on the exposed surfaces of certain epithelial cells.

ciliary body – a portion of the choroid of the eye that secretes aqueous humor and contains the ciliary muscle.

ciliates – protozoans that move by means of cilia.

circadian rhythm – a daily physiological or behavioral event, occurring on an approximate 24 hour cycle.

circumduction – a conelike movement of a body part, such that the distal end moves in a circle while the proximal portion remains relatively stable.

clitoris – a small, erectile structure in the vulva of the female.

cochlea – the spiral portion of the inner ear that contains the spiral organ (organ of Corti).

climax community – the final, stable stage in succession.

clone – asexually produced organisms having a consistent genetic constitution.

cnidarian – small aquatic organisms having radial symmetry and stinging cells with nematocysts.

cocoon – protective, or resting, stage of development in certain invertebrate animals.

codon – a "triplet" of three nucleotides in RNA that directs the placement of an amino acid into a polypeptide chain.

coelom – body cavity of higher animals, containing visceral organs.

collar cells – flagella-supporting cells in the inner layer of the wall of sponges.

colon – the first portion of the large intestine.

colony – an aggregation of organisms living together in close proximity.

common bile duct – a tube that is formed by the union of the hepatic duct and cystic duct, transports bile to the duodenum.

community – an ecological unit composed of all the populations of organisms living and interacting in a given area.

compact bone – tightly packed bone that is superficial to spongy bone; also called *dense bone*.

competition – interaction between individuals of the same or different species for a mutually necessary resource.

complete flower – a flower that has four whorls of floral components including sepals, petals, stamens, and carpels.

compound eye – arthropod eye consisting of multiple lenses.

compound leaf – a leaf blade divided into distinct leaflets.

condyle – a rounded process at the end of a long bone that forms an articulation.

conidia – spores produced by fungi during asexual reproduction.

conifer – a cone-bearing seed plant, such as pine, fir, and spruce.

conjugation – sexual union in which the nuclear material of one cell enters another.

connective tissue – one of the four basic tissue types within an animal's body. It is a binding and supportive tissue with abundant matrix.

consumer – an organism that derives nutrients by feeding upon another.

control – a sample in an experiment that undergoes all the steps in the experiment except the one being investigated.

convergent evolution – the evolution of similar structures in different groups of organisms exposed to similar environments.

coral – a cnidarian that has a calcium carbonate skeleton whose remains contribute to form reefs.

cork – the protective outer layer of bark of trees, composed of dead cells that may be sloughed off.

cornea – the transparent convex, anterior portion of the outer layer of the eye.

cortex – the outer layer of an organ such as the convoluted cerebrum, adrenal gland, or kidney.

costal cartilage – the cartilage that connects the ribs to the sternum.

cranial – pertaining to the cranium.

cranial nerve – one of twelve pairs of nerves that arise from the inferior surface of the brain.

cranium – the bones of the skull that enclose the brain and support the organs of sight, hearing, and balance.

crossing over – the exchange of corresponding chromatid segments of genetic material of homologous chromosomes during synapsis in meiosis I.

cuticle – waxlike covering on the epidermis of nonwoody plants to prevent water loss.

cyanobacteria – photosynthetic prokaryotes that have chlorophyll and release oxygen.

cytokinesis – division of the cellular cytoplasm.

cytology – the science dealing with the study of cells.

cytoplasm – the protoplasm of a cell located outside of the nucleus.

cytoskeleton – protein filaments throughout the cytoplasm of certain cells that help maintain the cell shape and provide movement.

deciduous – plants that seasonally shed their leaves.

dendrite – a nerve cell process that transmits impulses toward a neuron cell body.

denitrifying bacteria – single-cellular organisms of the kingdom Monera that convert nitrate to atmospheric nitrogen.

dentin – the principal substance of a tooth, covered by enamel over the crown and by cementum on the root.

dermis – the second, or deep, layer of skin beneath the epidermis.

descending colon – the segment of the large intestine that descends on the left side from the level of the spleen to the level of the left iliac crest.

detritus – non-living organic matter that is important in the nutrient cycle in soil formation.

diaphragm – a flat dome of muscle and connective tissue that separates the thoracic and abdominal cavities.

diaphysis – the shaft of a long bone.

diastole – the portion of the cardiac cycle during which the ventricular heart chamber wall is relaxed.

diarthrosis – a freely movable joint in a functional classification of joints.

diatoms – aquatic unicellular algae characterized by a cell wall composed of two silica impregnated valves.

dicot – a kind of angiosperm characterized by the presence of two cotyledons in the seed; also called *dicotyledon*.

diffusion – movement of molecules from an area of greater concentration to an area of lesser concentration.

dihybrid cross – a breeding experiment in which parental varieties differing in two traits are mated.

dimorphism – two distinct forms within a species, with regards to size, color, organ structure, and so on.

diphyodont – two sets of teeth, deciduous and permanent.

diploid – having two copies of each different chromosome, pairs of homologous chromosomes (2N).

distal – away from the midline or origin; the opposite of proximal.

division – a major taxonomic grouping of plants that includes classes sharing certain features with close biological relationships.

dominant – a hereditary characteristic that expresses itself even when the genotype is heterozygous.

dormancy – a period of suspended activity and growth.

dorsal – pertaining to the back or posterior portion of a body part; the opposite of ventral.

double helix – a double spiral used to describe the three-dimensional shape of DNA.

ductus deferens – a tube that carries spermatozoa from the epididymis to the ejaculatory duct; also called the *vas deferen*s or *seminal duct*.

duodenum – the first portion of the small intestine.

dura mater – the outermost meninx covering the central nervous system.

eccrine gland – a sweat gland that functions in body cooling.

ecology – the study of the relationship of organisms and the physical environment and their interactions.

ecosystem – a biological community and its associated abiotic environment.

ectoderm – the outermost of the three primary embryonic germ layers.

edema – an excessive retention of fluid in the body tissues.

effector – an organ such as a gland or muscle that responds to motor stimulation.

efferent – conveying away from the center of an organ or structure.

ejaculation – the discharge of semen from the male urethra during climax.

electrocardiogram – a recording of the electrical activity that accompanies the cardiac cycle; also called *ECG* or *EKG*.

electroencephalogram – a recording of the brain wave pattern; also called *EEG*.

electromyogram – a recording of the activity of a muscle during contraction; also called *EMG*.

electrolyte – a solution that conducts electricity by means of charged ions.

electron – the unit of negative electricity.

element – a structure comprised of only one type of atom (i.e., carbon, hydrogen, oxygen).

embryo – a plant or an animal at an early stage of development.

emulsification – the process of dispersing one liquid in another.

enamel – the outer, dense substance covering the crown of a tooth.

endocardium – the fibrous lining of the heart chambers and valves.

endochondral bone – bones that form as hyaline cartilage models first and then are ossified.

endocrine gland – a hormone-producing gland that secretes directly into the blood or body fluids.

endoderm – the innermost of the three primary germ layers of an embryo.

endodermis – a plant tissue composed of a single layer of cells that surrounds and regulates the passage of materials into the vascular cylinder of roots.

endometrium – the inner lining of the uterus.

endoskeleton – hardened, supportive internal tissue of echinoderms and vertebrates.

endosperm – a plant tissue of angiosperm or gymnosperm seeds that stores nutrients; the endosperm of angiosperms is 3n in chromosome number.

endothelium – the layer of epithelial tissue that forms the thin inner lining of blood vessels and heart chambers.

enzyme – a protein catalyst that activates a specific reaction.

eosinophil – a type of white blood cell that becomes stained by acidic eosin dye; constitutes about 2%–4% of the white blood cells.

epicardium – the thin, outer layer of the heart; also called the *visceral pericardium*.

epicotyl – plant embryo portion that contributes to stem development.

epidermis – the outermost layer of the skin, composed of stratified squamous epithelium.

epididymis – a coiled tube located along the posterior border of the testis—stores spermatozoa and discharges them during ejaculation.

epidural space – a space between the spinal dura mater and the bone of the vertebral canal.

epiglottis – a leaflike structure positioned on top of the larynx that covers the glottis during swallowing.

epinephrine – a hormone secreted from the adrenal medulla resulting in actions similar to those from sympathetic nervous system stimulation; also called *adrenaline*.

epiphyseal plate – a cartilaginous layer located between the epiphysis and diaphysis of a long bone, which functions in longitudinal bone growth.

epiphysis – the end segment of a long bone, distinct in early life but later becoming part of the larger bone.

epiphyte – nonparasitic plant, such as orchid and Spanish moss, that grows on the surface of other plants.

epithelial tissue – one of the four basic animal tissue types; the type of tissue that covers or lines all exposed body surfaces.

erection – a response within an organ, such as the penis, when it becomes turgid and erect as opposed to being flaccid.

erythrocyte – a red blood cell.

esophagus – a tubular organ of the GI tract that leads from the pharynx to the stomach.

estrogen – female sex hormone secreted from the ovarian (Graafian) follicle.

estuary – a zone of mixing between fresh water and seawater.

eukaryotic – possessing the membranous organelles characteristic of complex cells.

eustachian canal – see *auditory tube*.

evolution – genetic and phenotypic changes occurring in populations of organisms through time, generally resulting in increased adaptation for continued survival.

excretion – discharging waste material.

exocrine gland – a gland that secretes its product to an epithelial surface, directly or through ducts.

exoskeleton – an outer, hardened supporting structure secreted by ectoderm or epidermis.

expiration – the process of expelling air from the lungs through breathing out; also called *exhalation*.

extension – a movement that increases the angle between two bones of a joint.

external ear – the outer portion of the ear, consisting of the auricle (pinna), and the external auditory canal, and tympanum.

extracellular – outside a cell or cells.

extraembryonic membranes – membranes that are not a part of the embryo but are essential for the health and development of the organism.

extrinsic – pertaining to an outside or external origin.

facet – a small, smooth surface of a bone where articulation occurs.

facilitated transport – transfer of a particle into or out of a cell along a concentration gradient by a process requiring a carrier.

fallopian tube – see *uterine tube*.

fascia – a tough sheet of fibrous connective tissue binding the skin to underlying muscles or supporting and separating muscle.

fasciculus – a bundle of muscle or nerve fibers.

feces – waste material expelled from the GI tract during defecation, composed of food residue, bacteria, and secretions; also called *stool*.

fertilization – the fusion of two haploid gamete nuclei to form a diploid zygote nucleus.

fetus – the unborn offspring during the last stage of prenatal development.

fibrous root – an intertwining mass of many roots of about equal size.

filament – a long chain of cells.

filter feeder – an animal that obtains food by straining it from the water.

filtration – the passage of a liquid through a filter or a membrane.

fimbriae – fringelike extensions from the borders of the open end of the uterine tube.

fissure – a groove or narrow cleft that separates two parts of an organ.

flagella – long slender locomotor processes characteristic of flagellate protozoans, certain bacteria, and sperm.

flexion – a movement that decreases the angle between two bones of a joint; opposite of extension.

flora – a general term for plant life.

flower – the blossom of an angiosperm that contains the reproductive organs.

fluke –a parasitic flatworm within the class Trematoda.

follicle –the portion of the ovary that produces the egg and the female sex hormone, estrogen; the depression that supports and develops a feather or hair.

fontanel – a membranous-covered region on the skull of a fetus or baby where ossification has not yet occurred; also called a *soft spot*.

food web – the food links among populations in a community.

foot – the terminal portion of the lower extremity, consisting of the tarsus, metatarsus, and digits.

foramen – an opening in an anatomical structure for the passage of a blood vessel or a nerve.

foramen ovale – the opening through the interatrial septum of the fetal heart.

fossa – a depressed area, usually on a bone.

fossil – any preserved ancient remains or impressions of an organism.

fovea centralis – a depression on the macula lutea of the eye where only cones are located, which is the area of keenest vision.

frond – the leaf of a fern containing many leaflets.

fruit – a mature ovary enclosing a seed or seeds.

gallbladder – a pouchlike organ, attached to the inferior side of the liver, which stores and concentrates bile.

gamete – a haploid sex cell, sperm or egg.

gametophyte –the haploid, gamete-producing generation in the life cycle of a plant.

gamma globulins – proteins substances that act as antibodies often found in immune serums.

ganglion – an aggregation of nerve cell bodies outside the central nervous system.

gastrointestinal tract – the tubular portion of the digestive system that includes the stomach and the small and large intestines; also called the *GI tract*.

gene – part of the DNA molecule located in a definite position on a certain chromosome and coding for a specific product.

gene pool – the total of all the genes of the individuals in a population.

genetic drift – evolution by chance process.

genetics – the study of heredity.

genotype – the genetic makeup of an organism.

genus – the taxonomic category above species and below family.

geotropism – plant growth oriented with respect to gravity; stems grow upward, roots grow downward.

germ cells – gametes or the cells that five rise to gametes or other cells.

germination – the process by which a spore or seed ends dormancy and initiates normal metabolism, development, and growth.

gill – a gas exchange organ characteristic of fishes and other aquatic or semiaquatic animals.

gingiva – the fleshy covering over the mandible and maxilla through which the teeth protrude within the mouth; also called the *gum*.

girdling – removal of a strip of bark from around a tree down to the wood layer.

gland – an organ that produces a specific substance or secretion.

glans penis – the enlarged, distal end of the penis.

glomerular capsule – the double-walled proximal portion of a renal tubule that encloses the glomerulus of a nephron; also called *Bowman's capsule*.

glomerulus – a coiled tuft of capillaries that is surrounded by the glomerular capsule and filters urine from the blood.

glottis – a slit like opening into the larynx, positioned between the vocal folds.

glycogen – the principal storage carbohydrate in animals. It is stored primarily in the liver and is made available as glucose when needed by the body cells.

goblet cell – a unicellular gland within columnar epithelia that secretes mucus.

gonad – a reproductive organ, testis or ovary, that produces gametes and sex hormones.

granum – a "stack" of membrane flattened disks within the chloroplast that contain chlorophyll.

gray matter – the portion of the central nervous system that is composed of nonmyelinated nervous tissue.

grazer – an animal that feeds on low growing vegetation, such as grasses.

growth ring – a growth layer of secondary xylem (wood) or secondary phloem in gymnosperms or angiosperms.

guard cell – an epidermal cell to the side of a leaf stoma that helps to control the stoma size.

gut – pertaining to the intestine –generally a developmental term.

gymnosperm – a vascular seed-producing plant.

gyrus – a convoluted elevation or ridge.

habitat – the ecological abode of a particular organism.

hair – an epidermal structure consisting of keratinized dead cells that have been pushed up from a dividing basal layer.

hair cells – a specialized receptor nerve ending for responding to sensations, such as in the spiral organ of the inner ear.

hair follicle – a tubular depression in the skin in which a hair develops.

hand – the terminal portion of the upper extremity, consisting of the carpus, metacarpus, and digits.

haploid – having one copy of each different chromosome.

hard palate – the bony partition between the oral and nasal cavities, formed by the maxillae and palatine bones.

haversian system – see osteon

heart – a muscular, pumping organ positioned in the thoracic cavity.

hematocrit – the volume percentage of red blood cells in whole blood.

hemoglobin – the pigment of blood cells that transports O_2 and CO_2.

hemopoiesis – production of red blood cells.

hepatic portal circulation – the return of venous blood from the digestive organs and spleen through a capillary network within the liver before draining into the heart.

herbaceous – a nonwoody plant.

herbaceous stem – stem of a nonwoody plant.

herbivore – an organism that feeds exclusively on plants.

heredity – the transmission of certain characteristics, or traits, from parents to offspring, via the genes.

heterodont – having teeth differentiated into incisors, canines, premolars, and molars for specific functions.

heterotroph – an organism that utilizes preformed food.

heterozygous – having two different alleles (i.e., *Bb*) for a given trait.

hiatus – an opening or fissure.

hilum – a concave or depressed area where vessels or nerves enter or exit an organ.

histology – microscopic anatomy of the structure and function of tissues.

holdfast – basal extension of a multicellular alga that attaches it to a solid object.

homeostasis – a consistency and uniformity of the internal body environment which maintains normal body function.

homologous – similar in developmental origin and sharing a common ancestry.

homothalic – species in which individuals produce both male and female reproductive structures and are self fertile.

hormone – a chemical substance that is produced in an endocrine gland and secreted into the bloodstream to cause an effect in a specific target organ.

host – an organism on or in which another organism lives.

ileum – the terminal portion of the small intestine between the jejunum and cecum.

imprinting – a type of learned behavior occurring during a limited critical period.

indigenous – organisms that are native to a particular region; not introduced.

inguinal – pertaining to the groin region.

insertion – the more movable attachment of a muscle, usually more distal in location.

inspiration – the act of breathing air into the alveoli of the lungs; also called *inhalation*.

instar – stage of insect or other arthropod development between molts.

integument – pertaining to the skin.

internal ear – the innermost portion or chamber of the ear, containing the cochlea and the vestibular organs.

internode – region between stem nodes.

interstitial – pertaining to spaces of structure between the functioning active tissue of any organ.

intracellular – within the cell itself.

intervertebral disc – a pad of fibrocartilage between the bodies of adjacent vertebrae/

intestinal gland – a simple tubular digestive gland that opens onto the surface of the intestinal mucosa and secretes digestive enzymes; also called *crypt of Lieberkuhn*.

intrinsic – situated or pertaining to internal origin.

invertebrate – an animal that lacks a vertebral column.

iris – the pigmented vascular tunic portion of the eye that surrounds the pupil and regulates its diameter.

islets of Langerhans – see *pancreatic islets*.

isotope – a chemical element that has the same atomic number as another but a different atomic weight.

jejunum – the middle portion of the small intestine, located between the duodenum and the ileum.

joint capsule – the fibrous tissue that encloses the joint cavity of a synovial joint.

jugular – pertaining to the veins of the neck which drain the areas supplied by the carotid arteries.

karyotype – the arrangement of chromosomes that is characteristic of the species or of a certain individual.

keratin – an insoluble protein present in the epidermis and in epidermal derivatives such as hair and nails.

kidney – one of the paired organs of the urinary system that contains nephrons and filters wastes from the blood in the formation of urine.

kingdom – a taxonomic category grouping related divisions (plants) or phyla (animals).

labia major – a portion of the external genitalia of a female, consisting of two longitudinal folds of skin extending downward and backward from the mons pubis.

labia minora – two small folds of skin, devoid of hair and sweat glands, lying between the labia majora of the external genitalia of a female.

lacrimal gland – a tear-secreting gland, located on the superior lateral portion of the eyeball underneath the upper eyelid.

lactation – the production and secretion of milk by the mammary glands.

lacteal – a small lymphatic duct within a villus of the small intestine.

lacuna – a hollow chamber that houses an osteocyte in mature bone tissue or a chondrocyte in cartilage tissue.

lamella – a concentric ring of matrix surrounding the central canal in an osteon of mature bone tissue.

large intestine – the last major portion of the GI tract, consisting of the cecum, colon, rectum, and anal canal.

larva – an immature, developmental stage that is quite different from the adult.

larynx – the structure located between the pharynx and trachea that houses the vocal folds (chords); commonly called the *voice box*.

lateral root – a secondary root that arises by branching from an older root.

leaf veins – plant structures that contain the vascular tissues in a leaf.

legume – a member of the pea, or bean, family.

lens – a transparent refractive structure of the eye, derived from ectoderm and positioned posterior to the pupil and iris.

lenticel – spongy area in the bark of a stem or root that permits interchange of gases between internal tissues and the atmosphere.

leukocyte – a white blood cell; also spelled *leucocyte*.

lichen – algae or bacteria and fungi coexisting in a mutualistic relationship.

ligament – a fibrous band or cord of connective tissue that binds bone to one to strengthen and provide support to the joint; also may support viscera.

limbic system – a portion of the brain concerned with emotions and autonomic activity.

locus – the specific location or site of a gene within the chromosome.

lumbar – pertaining to the region of the loins.

lumen – the space within a tubular structure through which a substance passes.

lung – one of the two major organs of respiration within the thoracic cavity.

lymph – a clear fluid that flows through lymphatic vessels.

lymph node – a small, ovoid mass located along the course of lymph vessels.

lymphocyte – a type of white blood cell characterized by a granular cytoplasm.

macula lutea – a depression in the retina that contains the fovea centralis, the area of keenest vision.

malnutrition – any abnormal assimilation of food; receiving insufficient nutrients.

mammary gland – in mammals, the gland of the female breast responsible for lactation and nourishment of the young.

mantle – fleshy fold of the body wall of a mollusk, typically involved in shell formation.

marine – pertaining to the sea or ocean.

marrow –the soft vascular tissue that occupies the inner cavity of certain bones and produces blood cells.

matrix – the intercellular substance of a tissue.

mediastinum – the partition in the center of the thorax between the two pleural cavities.

medulla – the center portion of an organ.

medulla oblongata – a portion of the brain stem between the pons and the spinal cord.

medullary cavity – the hollow center of the diaphysis of a long bone, occupied by marrow.

megaspore – a plant spore that will germinate to become a female gametophyte.

meiosis – cell division by which haploid cells are formed from a diploid cell.

melanocyte – a pigment-producing cell in the deepest epidermal layer of the skin.

membranous bone – bone that forms from membranous connective tissue rather than from cartilage.

menarche – the first menstrual discharge.

meninges – a group of three fibrous membranes that cover the central nervous system.

meniscus – wedge-shaped cartilage in certain synovial joints.

menopause – the cessation of menstrual periods in the human female.

menses – the monthly flow of blood from the human female genital tract.

menstrual cycle – the rhythmic female reproductive cycle, characterized by changes in hormone levels and physical changes in the uterine lining.

menstruation – the discharge of blood and tissue from the uterus at the end of the menstrual cycle.

meristem tissue – undifferentiated plant tissue that is capable of dividing and producing new cells.

mesentery – a fold of peritoneal membrane that attaches an abdominal organ to the abdominal wall.

mesoderm – the middle one of the three primary germ layers.

mesophyll – the middle tissue layer of a leaf containing cells that are active in photosynthesis.

mesothelium – a simple squamous epithelial tissue that lines body cavities and covers visceral organs; also called *serosa*.

metabolism – the chemical changes that occur within a cell.

metacarpus – the region of the hand between the wrist and the digits, including the five bones that support the palm of the hand.

metamorphosis – change in morphologic form, such as when an insect larva develops into the adult or as a tadpole develops into an adult frog.

metatarsus – the region of the foot between the ankle and the digits containing five bones.

microbiology – the science dealing with microscopic organisms, including fungi, protozoa, and viruses.

microspore – a spore in seed plants that develops into a pollen grain, the male gametophyte.

microvilli – microscopic, hairlike projections of cell membranes on certain epithelial cells.

midbrain – the portion of the brain between the pons and the forebrain.

middle ear – the middle of the three ear chambers, containing the three auditory ossicles.

migration – movement of organisms from one geographical site to another.

mimicry – a protective resemblance of an organism to another.

mitosis – the process of cell division, in which the two daughter cells are identical and contain the same number of chromosomes.

mitral valve – the left atrioventricular heart valve; also called the *bicuspid valve*.

mixed nerve – a nerve containing both motor and sensory nerve fibers.

molecule – a minute mass of matter, composed of a combination of atoms that form a given chemical substance or compound.

molting – periodic shedding of an epidermal derived structure.

monocot – a type of angiosperm in which the seed has only a single cotyledon; also called *monocotyledon*.

motor neuron – a nerve cell that conducts an action potential away from the central nervous system and innervates effector organs (muscles and glands); also called *efferent neuron*.

motor unit – a single motor neuron and the muscle fibers it innervates.

mucosa – a mucous membrane that lines cavities and tracts opening to the exterior.

muscle – an organ adapted to contract, three types of muscle tissue are cardiac, smooth, and skeletal.

mutation – a variation in an inheritable characteristic, a permanent transmissible change in which the offspring differ from the parents.

mutualism – a beneficial relationship between two organisms of different species.

myelin – a lipoprotein material that forms a sheathlike covering around nerve fibers.

myocardium – the cardiac muscle layer of the heart.

myofibril – a bundle of contractile fibers within muscle cells.

myoneural junction – the site of contact between an axon of a motor neuron and a muscle fiber.

myosin – a thick filament protein that together with actin causes muscle contraction.

nail – a hardened, keratinized plate that develops form the epidermis and forms a protective covering on the digits.

nares – the opening into the nasal cavity; also called *nostrils*.

nasal cavity – a mucosa-lined space above the oral cavity, which is divided by a nasal septum and is the first chamber of the respiratory system.

natural selection – the evolutionary mechanism by which better adapted organisms are favored to reproduce and pass on their genes to the next generation.

nephron – the functional unit of the kidney, consisting of a glomerulus, glomerular capsule, convoluted tubules, and the nephron loop.

nerve – a bundle of nerve fibers outside the central nervous system.

neurofibril node – a gap in the myelin sheath of a nerve fiber; also called the *node of Ranvier*.

neuroglia – specialized supportive cells of the central nervous system.

neurolemmocyte – a specialized neuroglial cell that surrounds an axon fiber of a peripheral neuron and forms the neurolemmal sheath; also called the *Schwann cell*.

neuron – the structural and functional unit of the nervous system, composed of a cell body, dendrites, and an axon; also called a *nerve cell*.

neutron – a subatomic particle in the nucleus of an atom that has a weight of one atomic mass unit and carries no charge.

neutrophil – a type of phagocytic white blood cell.

niche – the position and functional role of an organism in its ecosystem.

nipple – a dark pigmented, rounded projection at the tip of the breast.

nitrogen fixation – a process carried out by certain organisms, such as by soil bacteria, whereby free atmospheric nitrogen is converted into ammonia compounds.

node – location on a stem where a leaf is attached.

node of Ranvier – see *neurofibril node*.

notochord – a flexible rod of tissue that extends the length of the back of an embryo.

nucleic acid – an organic molecule composed of joined nucleotides, such as RNA and DNA.

nucleus – a spheroid body within the eukaryotic cell that contains the chromosomes of the cell.

nut – a hardened and dry single-seeded fruit.

olfactory – pertaining to the sense of smell.

oocyte – a developing egg cell.

oogenesis – the process of female gamete formation.

oogonium – a unicellular female reproductive organ of various protists that contains a single or several eggs.

optic – pertaining to the eye and the sense of vision.

optic chiasma – an X-shaped structure on the inferior aspect of the brain where there is a partial crossing over of fibers in the optic nerves.

optic disc – a small region of the retina where the fibers of the ganglion neurons exit from the eyeball to form the optic nerve; also called the *blind spot*.

oral – pertaining to the mouth; also called *buccal*.

organ – a structure consisting of two or more tissues, which perform a specific function.

organelle – a minute structure of the eukaryotic cell that performs a specific function.

organism – an individual living creature.

orifice – an opening into a body cavity or tube.

origin – the place of muscle attachment onto the more stationary point or proximal bone; opposite the insertion.

osmosis – the diffusion of water from a solution of lesser concentration to one of greater concentration through a semipermeable membrane.

ossicle – one of the three bones of the middle ear.

osteocyte – a mature bone cell.

osteon – a group of osteocytes and concentric lamellae surrounding a central canal within bone tissue; also called a *haversian system*.

osteon – a group of osteocytes and concentric lamellae surrounding a central canal within one tissue; also called a *haversian system*

oval window – See *vestibular window*.

ovarian follicle – a developing ovum and its surrounding epithelial cells.

ovary – the female gonad in which ova and certain sexual hormones are produced.

oviduct – the tube that transports ova from the ovary to the uterus; also called the **uterine tube** or **fallopian tube**.

ovipositor – a structure at the posterior end of the abdomen in many female insects for laying eggs.

ovulation – the rupture of an ovarian follicle with the release of an ovum.

ovule – the female reproductive structure in a seed plant that contains the megasporangium where meiosis occurs and the female gametophyte is produced.

ovum – a secondary oocyte after ovulation but before fertilization.

palisade layer – the upper layer of the mesophyll of a leaf that carries out photosynthesis.

pancreas – organ in the abdominal cavity that secretes gastric juices into the GI tract and insulin and glucagon into the blood.

pancreatic islets – a cluster of cells within the pancreas that forms the endocrine portion of the pancreas; also called *islets of Langerhans*.

papillae – small nipplelike projections.

paranasal sinus – an air chamber lined with a mucous membrane that communicates with the nasal cavity.

parasite – an organism that resides in or on another from which it derives sustenance.

parasympathetic – pertaining to the division of the autonomic nervous system concerned with activities that are antagonistic to the sympathetic division of the autonomic nervous system.

parathyroids – small endocrine glands that are embedded on the posterior surface of the thyroid glands and are concerned with calcium metabolism.

parenchyma – the principal structural cells of plants.

parietal – pertaining to a wall of an organ or cavity.

parotid gland – one of the paired salivary glands on the sides of the face over the masseter muscle.

parturition – the process of childbirth.

pathogen – any disease-producing organism.

pectin – an organic compound in the intercellular layer and primary wall of plant cell walls; the basis of fruit jellies.

pedicel – the stalk of a flower in an inflorescence.

pectoral girdle – the portion of the skeleton that supports the upper extremities.

pelvic – pertaining to the pelvis.

pelvic girdle – the portion of the skeleton to which the lower extremities are attached.

penis – the external male genital organ, through which urine passes during urination and which transports semen to the female during coitus.

perennial – a plant that lives through several to many growing seasons.

pericardium – a protective serous membrane that surrounds the heart.

pericarp – the fruit wall that forms from the wall of a mature ovary.

perineum – the floor of the pelvis.

periosteum – a fibrous connective tissue covering the outer surface of bone.

peripheral nervous system – the nerves and ganglia of the nervous system that lie outside of the brain and spinal cord.

peristalsis – rhythmic contractions of smooth muscle in the walls of various tubular organs, which move the contents along.

peritoneum – the serous membrane that lines the abdominal cavity and covers the abdominal viscera.

petal – the leaf of a flower, which is generally colored.

petiole – structure of a leaf that connects the blade to the stem.

phagocyte – any cell that engulfs other cells, including bacteria, or small foreign particles.

phalanx – a bone of the finger or toe.

pharynx – the organ of the GI tract and respiratory system located at the back of the oral and nasal cavities and extending to the larynx anteriorly and the esophagus posteriorly; also called the *throat*.

phenotype – the appearance of an organism caused by the genotype and environmental influences.

pheromone – a chemical secreted by one organism that influences the behavior of another.

phloem – vascular tissue in plants that transports nutrients.

photoperiodism – the response of an organism to periods of light and dark.

photosynthesis – the process of using the energy of the sun to make carbohydrates from carbon dioxide and water.

phototropism – plant growth or movement in response to a directional light source.

physiology – the science that deals with the study of body functions.

phytoplankton – microscopic, free-floating, photosynthetic organisms that are the major producers in freshwater and marine ecosystems.

pia mater – the innermost meninx that is in direct contact with the brain and spinal cord.

pineal gland – a small, cone-shaped gland located in the roof of the third ventricle of the brain.

pistil – the reproductive structure of a flower that is composed of the stigma, style, and ovary.

pith – a centrally located tissue within a dicot stem.

pituitary gland – a small, pea-shaped endocrine gland situated on the inferior surface of the brain that secretes a number of hormones; also called the *hypophysis*.

placenta – the organ of metabolic exchange between the mother and the fetus.

plankton – aquatic, free-floating microscopic organisms.

plasma – the fluid, extracellular portion of circulating blood.

plastid – an organelle of a plant that assists chloroplasts in photosynthesis.

platelets – fragments of specific bone marrow cells that function in blood coagulation; also called *thrombocytes*.

pleural membranes – serous membranes that surround the lungs and line the thoracic cavity.

plexus – a network of interlaced nerves or vessels.

pollen grain – male gametophyte generation of seed plants.

pollination – the delivery by wind, water, or animals of pollen to the stigma of a pistil in flowering plants, leading to fertilization.

polypeptide – a molecule of many amino acids linked by peptide bonds.

pons – the portion of the brain stem just above the medulla oblongata and anterior to the cerebellum.

population – all the organisms of the same species in a particular location.

posterior (*dorsal*) – toward the back.

predation – the consumption of one organism by another.

pregnancy – a condition where a female has a developing offspring in the uterus.

prenatal – the period of offspring development during pregnancy; before birth.

prey – organisms that are food for a predator.

producers – organisms within an ecosystem that synthesize organic compounds form inorganic constituents.

prokaryote – organism, such as a bacterium, that lacks the specialized organelles characteristic of complex cells.

proprioceptor – a sensory nerve ending that responds to changes in tension in a muscle or tendon.

prostate – a walnut-shaped gland surrounding the male urethra just below the urinary bladder that secretes an additive to seminal fluid during ejaculation.

protein – a macramolecule composed of one or several polypeptides.

prothallus – a heart-shaped structure that is the gametophyte generation of a fern.

proton – a subatomic particle of the atom nucleus that has a weight of one atomic mass unit and carries a positive charge; also a *hydrogen ion*.

proximal – closer to the midline of the body or origin of an appendage; opposite of distal.

puberty – the period of development in which the reproductive organs become functional.

pulmonary – pertaining to the lungs.

pupil – the opening through the iris that permits light to pass through the lens and enter the posterior cavity.

radial symmetry – symmetry around a central axis to that any half of an organism is identical to the other.

receptacle – the tip of the axis of a flow stalk that bears the floral organs.

receptor – a sense organ or a specialized end of a sensory neuron that receives stimuli from the environment.

rectum – the terminal portion of the GI tract, between the sigmoid colon and the anal canal.

reflex arc – the basic conduction pathway through the nervous system, consisting of a sensory neuron, association neuron, and a motor neuron.

regeneration – regrowth of tissue or the formation of a complete organism from a portion.

renal – pertaining to the kidney.

renal corpuscle – the portion of the nephron consisting of the glomerulus and a glomerular capsule.

renal pelvis – the inner cavity of the kidney formed by the expanded ureter and into which the calyces open.

renewable resource – a commodity that is not used up because it is continually produced in the environment.

replication – the process of producing a duplicate; a copying or duplication, such as DNA replication.

respiration – the exchange of gases between the external environment and the cells of an organism; the metabolic activity of cells resulting in the production of ATP.

retina – the inner layer of the eye that contains the rods and cones.

rhizome – an underground stem in some plants that stores photosynthetic products and gives rise to above-ground stems and leaves.

rod – a photoreceptor in the retina of the eye that is specialized for colorless, dim light vision.

root – the anchoring subterranean portion of a plant that permits absorption and conduction of water and mineral.

root cap – end mass of parenchyma cells which protects the apical meristem of a root.

root hair – unicellular epidermal projection from the root of a plant, which functions in absorption.

rugae – the folds or ridges of the mucosa of an organ.

sagittal – a vertical plane through the body that divides it into right and left portions.

salinity – saltiness in water or soil; a measure of the concentration of dissolved salts.

salivary gland – an accessory digestive gland that secretes saliva into the oral cavity.

sarcolemma – the cell membrane of a muscle fiber.

sarcomere – the portion of a skeletal muscle fiber between the two adjacent Z lines that is considered the functional unit of a myofibril.

savanna – open grassland with scattered trees.

Schwann cell – see *neurolemmocyte*.

sclera – the outer white layer of connective tissue that forms the protective covering of the eye.

sclerenchyma – supporting tissue in plants composed of hollow cells with thickened walls.

scolex – head region of a tapeworm.

scrotum – a pouch of skin that contains the testes and their accessory organs.

sebaceous gland – an exocrine gland of the skin that secretes sebum, an oily protective product.

secondary growth – plant growth in girth from secondary or lateral meristems.

seed – a plant embryo with a food reserve that is enclosed in a protective seed coat; seeds develop from mature ovules.

semen – the secretion of the reproductive organs of the male, consisting of spermatozoa and additives.

semilunar valve – crescent-shaped heart valves, positioned at the entrances to the aorta and the pulmonary trunk.

sensory neuron – a nerve cell that conducts an impulse from a receptor organ to the central nervous system – also called *afferent neurons*.

sepal – outermost structure of a flower beneath the petal; collectively called the *calyx*.

serous membrane – an epithelial and connective tissue membrane that lines body cavities and covers viscera; also called *serosa*.

sesamoid bone – a membranous bone formed in a tendon in response to joint stress.

sessile – organisms that lack locomotion and remain stationary, such as sponges and plants.

shoot – portion of a vascular plant that includes a stem with its branches and leaves.

sinoatrial node – a mass of cardiac tissue in the wall of the right atrium that initiates the cardiac cycle; the *SA node*; also called the *pacemaker*.

sinus – a cavity or hollow space within a body organ such as a bone.

skeletal muscle – a type of muscle tissue that is multinucleated, occurs in bundles, has crossbands of proteins, and contract either in a voluntary or involuntary fashion.

small intestine – the portion of the GI tract between the stomach and the cecum, functions in absorption of food nutrients.

smooth muscle – a type of muscle tissue that is nonstriated, composed of fusiform, single-nucleated fibers, and contracts in an involuntary, rhythmic fashion within the walls of visceral organs.

solute – a substance dissolved in a solvent to form a solution.

solvent – a fluid such as water that dissolves solutes.

somatic – pertaining to the nonvisceral parts of the body.

somatic cells – all the cells of the body of an organism except the germ cells.

sorus – a cluster of sporangia on the underside of fern pinnae.

species – a group of morphologically similar (common gene pool) organisms that are capable of interbreeding and producing fertile offspring and are reproductively isolated.

spermatic cord – the structure of the male reproductive system composed of the ductus deferens, spermatic vessels, nerve, cremasteric muscle, and connective tissue.

spermatogenesis – the production of male sex gametes, or spermatozoa.

spermatozoon – a sperm cells, or gamete.

sphincter – a circular muscle that constricts a body opening or the lumen of a tubular structure.

spinal cord – the portion of the central nervous system that extends from the brain stem through the vertebral canal.

spinal nerve – one of the thirty-one pairs of nerves that arise from the spinal cord.

spiracle – a respiratory opening in certain animals such as arthropods and sharks.

spirillum – a spiral-shaped bacterium.

spleen – a large, blood-filled organ located in the upper left of the abdomen and attached by the mesenteries to the stomach.

spongy bone – a type of bone that contains many porous spaces; also called *cancellous bone*.

spore – a reproductive cell capable of developing into an adult organism without fusion with another cells.

sporangium – an organ within which spores are produced.

sporophyll – a sporangium-bearing leaf.

stamen – the structure of a flower which is composed of a filament and an anther, where pollen grains are produced.

starch – carbohydrate molecule synthesized from photosynthetic products; common food storage substance in plants.

stigma – the upper portion of the pistil of a flower.

stele – the vascular tissue and pith or ground tissue at the central core of a root or stem.

stoma – an opening in a plant leaf through which gas exchange takes place.

stomach – a pouchlike digestive organ between the esophagus and the duodenum.

style – the long slender portion of the pistil of a flower.

submucosa – a layer of supportive connective tissue that underlies a mucous membrane.

succession – the sequence of ecological stages by which a particular biotic community gradually changes until there is a community of climax vegetation.

sucrose – a disaccharide (double sugar) consisting of a linked glucose and fructose molecule; the principal transport sugar in plants.

surfactant – a substance produced by the lungs that decreases the surface tension within the alveoli.

suture – a type of fibrous joint articulating between bones of the skull.

symbiosis – a close association between two organisms where one or both species derive benefit.

sympathetic – pertaining to that part of the autonomic nervous system concerned with activities antagonistic to the parasympathetic.

synapse – a minute space between the axon terminal of a presynaptic neuron and a dendrite of a postsynaptic neuron.

syngamy – union of gametes in sexual reproduction; fertilization.

synovial cavity – a space between the two bones of a synovial joint, filled with synovial fluid.

system – a group of body organs that function together.

systole – the muscular contraction of the ventricles of the heart during the cardiac cycle.

systolic pressure – arterial blood pressure during the ventricular systolic phase of the cardiac cycle.

taproot – a plant root system in which a single root is thick and straight.

target organ – the specific body organ that a particular hormone affects.

tarsus – pertaining to the ankle; the proximal portion of the foot that contains the tarsal bones.

taxonomy – the science of describing, classifying, and naming organisms.

tendo calcaneous – the tendon that attaches the calf muscles to the calcaneous bone.

tendon – a band of dense regular connective tissue that attaches muscle to bone.

testis – the primary reproductive organ of a male, which produces spermatozoa and male sex hormones.

tetrapod – a four-appendaged vertebrate, such as amphibian, reptile, bird, or mammal.

thoracic – pertaining to the chest region.

thorax – the chest.

thymus gland – a bilobed lymphoid organ positioned in the upper mediastinum, posterior to the sternum and between the lungs.

tissue – an aggregation of similar cells and their binding intercellular substance, joined to perform a specific function.

tongue – a protrusible muscular organ on the floor of the oral cavity.

toxin – a poisonous compound.

trachea – a tubule in the respiratory system of some invertebrates; the airway leading from the larynx to the bronchi in the respiratory system of vertebrates; also called the *windpipe*.

tract – a bundle of nerve fibers within the central nervous system.

trait – a distinguishing feature studied in heredity.

transpiration – the evaporation of water from a leaf, which pulls water from the roots through the stem to the leaf.

tricuspid valve – the heart valve between the right atrium and the right ventricle.

turgor pressure – osmotic pressure that provides rigidity to a cell.

tympanic membrane – the membranous eardrum positioned between the outer and middle ear; also called the *tympanum*, or the *ear drum*.

umbilical cord – a cordlike structure containing the umbilical arteries and vein, which connects the fetus with the placenta.

umbilicus – the site where the umbilical cord was attached to the fetus; also called the navel.

ureter – a tube that transports urine from the kidney to the urinary bladder.

urethra – a tube that transports urine from the urinary bladder to the outside of the body.

urinary bladder – a distensible sac in the pelvic cavity which stores urine.

uterine tube – the tube through which the ovum is transported to the uterus and where fertilization takes place; also called the *oviduct* or *fallopian tube*.

uterus – a hollow, muscular organ in which a fetus develops. It is located within the female pelvis between the urinary bladder and the rectum.

uvula – a fleshy, pendulous portion of the soft palate that blocks the nasopharynx during swallowing.

vacuole – a fluid-filled organelle.

vagina – a tubular organ that leads from the uterus to the vestibule of the female reproductive tract and receives the penis during coitus.

vascular tissue – plant tissue composed of xylem and phloem, functioning in transport of water, nutrient, and photosynthetic products throughout the plant.

vegetative – plant parts not specialized for reproduction; asexual reproduction.

vein – a blood vessel that conveys blood toward the heart.

ventral (*anterior*) – toward the front surface of the body.

vertebrate – an animal that possesses a vertebral column.

vestibular folds – the supporting folds of tissue for the vocal folds within the larynx.

vestibular window – a membrane-covered opening in the bony wall between the middle and inner ear, into which the footplate of the stapes fits; also called *oval window*.

viscera – the organs within the abdominal or thoracic cavities.

vitreous humor – the transparent gel that occupies the space between the lens and retina of the eye.

vocal folds – folds of the mucous membrane in the larynx that produce sound as they are pulled taut and vibrated; also called *vocal cords*.

vulva – the external genitalia of the female that surround the opening of the vagina; also called the *pudendum*.

wood – interior tissue of a tree composed of secondary xylem.

zoospore – a flagellated or ciliated spore produced asexually by some protists.

zygote – a fertilized egg cell formed by the union of a sperm and an ovum.

Index